Bicycling
Science

Bicycling Science

second edition

Frank Rowland Whitt

David Gordon Wilson

The MIT Press
Cambridge, Massachusetts
London, England

Third printing, 1985
© 1982 by The Massachusetts Institute of Technology

This book was set in Melior by DEKR Corporation and printed and bound by Murray Printing Co. in the United States of America.

Library of Congress Cataloging in Publication Data

Whitt, Frank Rowland.
    Bicycling science.

    Includes bibliographies and index.
    1. Bicycles—Dynamics. 2. Man-machine systems.
I. Wilson, David Gordon, 1928–.
II. Title.
TL410.W48 1982 629.2′31 81-20950
ISBN 0-262-23111-5   AACR2
ISBN 0-262-73060-X (pbk.)

Contents

Foreword

In Chest Springs, a rural town in Pennsylvania, a worn and rusting replica of a high-wheeler bicycle stands at the end of a lane. From the handlebars swings an equally aged mailbox. The postman enjoys delivering to that address because the bike reminds him of history. And he is not alone; many Americans think of the bike "historically" as a monument to an age gone by.

The bicycle has indeed enjoyed a rich and diverse history, culminating in the dizzying number of inventions that brought the safety bicycle, from which the machines we ride today are not radically different. But the bicycle was just the first stop on a journey that would take much of the world into the machine age. It was quite natural for gifted bike mechanics and inventors such as the Wright brothers and Henry Ford to apply their inventiveness to other vehicles. Such is progress.

One of the unfortunate by-products of progress was the declining interest in the bicycle. After all, motorcycles and cars went much faster and made much more noise. Equally depressing was the fact that entire generations of Americans believed that the bicycle was a child's toy. No wonder so many American soldiers were amazed at the number of bicycles flooding the roads of Europe.

It is generally agreed that the 1860s saw the development of the bicycle as we have come to know it today. During that decade, rubber was used for the first time to cushion the ride and ball bearings were introduced to provide easier pedaling and steering. Almost a hundred years later, America experienced the early signs of another bicycle revolution. Sports bikes with multiple gearing were introduced into the adult market. Cycling was promoted as an adult activity, as a legitimate life sport that would foster cardiovascular health. A vital connection had been made.

This second bicycle revolution gives every indication of being broad-based, deep, and diverse. Millions of people are riding bikes for exercise and transportation, and the market is alive with inventiveness. Large and small manufacturers are introducing new bikes, components, and systems at a rapid rate. We are witnessing high interest in aerodynamics, human power, and optimum riding position.

Because of the renewed interest in bicycles among consumers, students, and engineers, the publication of the second edition of *Bicycling Science*—which is really a new book with excerpts from the first—is very timely indeed. *Bicycling Science,* written by two distinguished engineers and practicing cyclists, provides a full and rich treatment of the physics and physiology of cycling. Although cycling is a wondrously simple affair, it involves physical phenomena such as braking, steering, rolling resistance, and stress. The strength and beauty of *Bicycling Science* is that the cycling is never lost for the physics. This is a very readable book.

Anyone involved in the bicycle industry should make it a point to read this book. The invaluable information on bike and component design will dispel many of the prevailing myths. But the book is not only for specialists. Anyone who takes more than passing pleasure in cycling would enjoy it—particularly the second part, entitled "Some bicycle physics."

Bicycling Science is richly educational, a basic tool for teaching elementary physics and physiology at all levels. A curriculum could be built around it.

Perhaps a final test of a book's worth is whether or not it "moves men." I think *Bicycling Science* will. The chapters dealing with

innovative transmissions, alternative forms of
the bicycle, and future possibilities for human-
powered vehicles offer a challenge to those who
will build the bikes of the future. Whitt and
Wilson have opened up the bicycle to another
age of discovery.

James C. McCullagh
Editor and Publisher, *Bicycling* magazine

Preface

We intended the first edition of this book "to be of interest to all mechanically inquisitive bicyclists, as well as to teachers of elementary mechanics or physiology, and to engineers and others working on approaches to lessen our dependence on high-energy-consumption transportation." Since we wrote those words, in 1974, several developments have seemed to confirm that the wave of popularity of the bicycle that started in the early 1970s was not just a short-term craze. The forced rise in world oil prices and the occasional disappearance of easily available gasoline (more significant in the United States) at last convinced many people that a long-term change was required in the affluent way of life experienced by many in the technologically advanced countries. Bicycling began to be taken more seriously as an alternative to the use of the automobile and public transit for commuting.

Another development with strong effects— good and bad—was the New York City transit strike of 1980. On the good side was the discovery by tens of thousands of people that commuting by bicycle was possible, and by many that it was pleasant. On the negative side, there were many accidents between bicyclists and pedestrians and between motor vehicles and bicycles, partly because of nonexistent or ill-conceived traffic arrangements for bicycles.

A third development, wholly beneficial, was the creation in 1974–75 of a new class of cycle racing. The International Human-Powered Vehicle Association, formed by Chester Kyle and a small group of fellow enthusiasts in California, sponsors speed trials and other races in which there are no restrictions on vehicle design other than that there must be no energy storage. The speeds already reached by the application of sophisticated aerodynamic fairings and supine or recumbent riding positions alone would have seemed incredible a decade ago, and yet it seems likely that 30 m/sec (67 mph) will be attained within a decade. Enthusiasm for this new sport is spreading and growing in the United States and in Europe. The attendance at the first racing meet in Britain in 1980 was more than for all the previous U.S. meets combined. This form of racing is certain to bring about a resumption in the development of bicycles for everyday use. A stream of new ideas was encouraged by bicycle racing in the 1865–1895 period, but this stream was then reduced to a trickle by the adoption of highly restrictive rules for racing. Now we see new developments in bicycle technology coming almost as a flood.

These developments have been largely responsible for this second edition of *Bicycling Science*. We have added a large amount of new information about human power output under various conditions, and have revised and expanded the sections on aerodynamic, wheel, and bearing losses. These inputs and outputs have been combined in a new chapter on the prediction of speeds for typical and hypothetical vehicles for various levels of power input. Thus we have tried to serve the new wave of designers, planners, and builders of vehicles both for racing and for everyday commuting use with data and methods that should further the designing of optimum vehicles.

We have also added a short chapter on the technological history of bicycles and tricycles, partly because it is a fascinating story and partly because awareness of what has been tried before can help to preclude the repetition of expensive mistakes. In this respect we have the same aim as "Professor" Archibald Sharp (who was in fact an instructor in engineering design at a London technical college), who wrote his classic *Bicycles and Tricycles* at a time (1896) when, as at present, people were experimenting with all manner of variations of cycle design and construction. In his preface Sharp wrote that "there are many frames on the market which evince on the part of their designers utter ignorance of mechanical science," and that "if the present work is the means of influencing makers, or purchasers, to such an extent as to make the manufacture and sale of such mechanical monstrosities in the future more difficult than it has been in the past, the author will regard his labors as having been entirely successful."

Other good books on the science of bicycling were published by authors such as R. P. Scott and C. Bourlet in the same period. From that time until the present revival of interest in bicycling, technical authors turned their attention toward automobiles, airplanes, and other apparently more exciting challenges. The stagnation of bicycle design, brought about largely by restrictive rules for racing, was aided by the lack of interest of publishers (and, perhaps, potential readers) and by the astonishing new transportation competitors—subways, cable and electric streetcars, motorcycles, automobiles, the railroads then reaching over 100 mph (about 50 m/sec), airships, and the early aircraft. We point out in the first chapter that a similar, though shorter, period of stagnation occurred after 1825, and that this was probably due to somewhat similar excitement about the potential of railroad transportation. Inventive people making improved bicycles in such periods of stagnation found that their concepts (and their manuscripts) fell on stony ground.

We as authors and bicyclists are fortunate to be living at a time when bicycle design is undergoing considerable change. In providing a technical guide, we have tried to start at all times from basic principles—which are, in general, the laws of physics. We have been concerned principally with dynamics rather than with stat-

ics. We have given raw data in those many
cases where the final answer, if there ever is
such a conclusion to research, is not yet known.
And occasionally we have made our own
estimates.

Some readers may be interested to learn how
this book came to be written. Frank Whitt, who
is a chemical engineer, had been a contributor
to (and for a period the technical editor of)
Cycle Touring (Cyclists' Touring Club, U.K.)
and had contributed technical papers to sympo-
sia and articles to magazines such as *Bicycling.*
He put these together into the beginnings of a
book. David Wilson was teaching mechanical-
engineering design at the Massachusetts Institute of Technology, using bicycles as occasional
examples and supervising some undergraduate
projects and theses. He had in Britain a small
savings account which the Bank of England
would not allow to be transferred to the United
States. With the help of the journal *Engineering*,
and with prize money from the savings account
and a contribution from Liberty Mutual Insur-
ance, he organized in 1967 an international
competition for developments in human-powered transportation. Whitt was one of the 73
entrants. They met some time after the competition was completed in 1969. Subsequently,
Whitt asked Wilson if he could find an American publisher for his manuscript. He had not
been successful in this endeavor in Britain,
and Wilson at first did no better in the United
States. Publishers felt that, whatever the quality
of a book on bicycling science, the potential
readership was so small that the considerable
exenditure of publishing the book was not
justified.

Then came the 1970s and the revival of interest in bicycling. There was still no sign of any
change in bicycle design, but Frank Satlow of
the MIT Press decided to take a long shot by
proposing that the book be adopted. The manuscript was accepted on the condition that Wil-

son add to it the results of the 1967–69 design competition and any relevant research data, and edit the whole book. That first edition was published in hard cover in 1975 and in paperback in 1977.

The continuing popularity of bicycling since then, the wealth of new developments and data, and in particular the intense interest in new types of vehicles made us wish almost immediately that we could rewrite the book. We were, therefore, delighted when Frank Satlow asked us if we would like to work on a second edition. As intimated above, although this is called a second edition, it is really a new book in scope and style; we hope that it will be received with the same goodwill and grace as was the first.

This preface is being written, sadly, by David Wilson alone. Frank Whitt suffered a paralyzing stroke in mid-1981, and as of the time of writing (September 1981) he has not yet been able to talk or to write. He is making slow progress, and it is hoped that he will be back with his insights, his experimental and design skills, and his wealth of information to contribute to us all. He is greatly missed.

Acknowledgments

Many individuals and organizations have helped to make this book possible.

Those who have given permission to reproduce copyrighted illustrations are acknowledged in the legends and are remembered here with appreciation.

William A. Bush, Evelyn Beaumont, Vaughan Thomas, Derek Roberts, and David E. Twitchett and other members of the Southern Veteran Cycle Club (U.K.) and the Camden Historical Society have lent items of equipment for testing and pieces of historical literature for study and copying.

Allen Armstrong, the superb mechanical designer responsible for the Positech derailleur shifting system and the dual-leverage brake, gave useful test data and photographs.

Fred DeLong, technical editor of several bicycling publications and author of *DeLong's Guide to Bicycles and Bicycling*, dedicates his free time to the cause of bicycling safety and has provided us with much helpful information.

Günter Fieblinger, professor at the University of Kassel, who helped to organize a highly successful bicycling congress in Bremen in 1980 ("Velo-City"), translated the graphs from the first edition of *Bicycling Science* into S.I. units for his students and gave us copies.

Richard Forrestall and Harald Maciejewski, designers and engineering perfectionists, formed FOMAC, Inc. to develop and produce recumbent bicycles, and have provided us with much valuable input.

Keith Hutcheon, technical director of T. I. Raleigh, Ltd., has provided helpful data on new products (particularly braking systems), historical information, and illustrations.

Chester Kyle, professor of mechanical engineering at Long Beach, California, is the one person most responsible for the founding of the International Human-Powered Vehicle Association and for the new generation of streamlined fast human-powered vehicles. He has given us a wealth of data, research papers, photographs, and encouragement.

Hans-Erhard Lessing, professor at the University of Ulm and author of *Das Fahrradbuch* (The Bicycle Book), has sent us valuable historical and scientific data about the development of the bicycle in Germany and of ergonomic research.

James C. (Chuck) McCullagh, editor of *Bicycling,* must be recognized here for his committed support of all that is good in present bicycling and of new developments.

Len Phillips, senior editor of *Technology Review* and avid bicyclist and photographer, fed us news, illustrations, and enthusiasm.

Anna Piccolo deciphered our rough typing and rougher scribbling over many months, as she did for the first edition, and cheerfuly preserved our sanity and hers at times when other demands were strident.

H. John Way, editor of *Cycle Touring,* has allowed the use of a considerable number of articles contributed to that magazine over the years by the senior author.

David Wilson's long-suffering family, Erica Sears Wilson, John M. B. Wilson, and Anne Sears Wilson, put up with taking second place to "BS" with grace, and have welcomed him back.

Note on Units

We have given values in S.I. (Système International) units in addition to those more familiar to English-speaking readers. Where a measure is referred to repeatedly (for instance, a one-inch-diameter tube) we have generally given the S.I. equivalent (25.4 mm) at the first mention only. We have not always translated historical measures. Other instances where we may have been inconsistent have been wheel and tire sizes and gearing, none of which translates directly by standard conversion factors. We have tried to explain such cases in the text.

HUMAN POWER

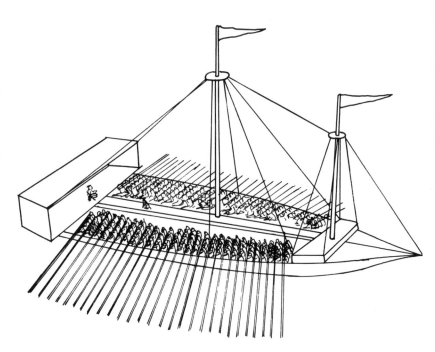

Figure 1.1
Early-seventeenth-
century galley, with
drummer in stern and
overseer on central
gangway. From a
drawing in the British
Museum, reproduced in
the *Encyclopaedia
Brittanica.*

History of human-powered machines and vehicles

It was through the use of tools that human beings raised themselves above the animals. In the broadest sense of the term, a tool might be something as simple as a stone hammer or as complex as a computer controlling a spacecraft. We are concerned with the historical and mechanical range of tools that led to the bicycle, which—almost alone among major human-powered machines—came to use human muscles in a near-optimum way. A short review of the misuse of human muscle power throughout history shows the bicycle to be a brilliant culmination of the efforts of many people to end such drudgery.

Many boats, even large ones, were muscle-powered until the seventeenth century. Roman galleys had hundreds of sweeps in up to three banks. Figure 1.1 shows a large seventeenth-century galley having 54 sweeps, with five men on each. The men were likely to be criminals, chained to their benches. A central gangway was patrolled by overseers equipped to provide persuasion for anyone considered to be taking life too easy. The muscle actions used by these unfortunate oarsmen were typical of those considered appropriate in the ancient world. The hand, arm, and back muscles were used the most, while the largest muscles in the body—those in the legs—were used merely to provide props or reaction forces. The motion was generally one of straining mightily against a slowly yielding resistance. With five men on the inboard end of a sweep, the one at the extreme end would have a more rapid motion than the one nearest to the pivot, but even the end man would probably be working at well below his optimum speed.

Figure 1.2
Engraving showing use
of capstans in erection of
an obelisk at the Vatican
in 1586. (The penalty for
disrupting work was
death.) From N. Zabaglia,
Castelli e Ponti (Rome,
1743).

Most farm work and forestry fell into the same
general category. Hoeing, digging, sawing,
chopping, pitchforking, and shoveling all used
predominantly the arm and back muscles, with
little useful output from the leg muscles. In
many cases, the muscles had to strain against
stiff resistances; it is now known that muscles
are most efficient and develop maximum power
when they are contracting quickly against a
small resistance, in a good "impedance match."

One medieval example of the use of appropriate muscles in a good impedance match is the
capstan (figure 1.2). Several people walked in a
circle, pushing on a radial arms, to winch in a
rope. The capstan's diameter was chosen to give
comfortable working conditions, and each
pusher could choose the preferred radial position on the bar.

Other relatively satisfactory uses of muscle

Figure 1.3
Inclined footmill.
Reproduced, with
permission, from Aubrey
F. Burstall, *A History of
Mechanical Engineering*
(London: Faber & Faber,
1963).

power were the inclined treadmill (figure 1.3),
Leonardo da Vinci's drum or cage for armaments (rotated by people climbing on the outside),[1] and treadmill-driven pumps (figure 1.4).
This type of work may not have been pleasant,
but per unit of output it was far more congenial
than that of a galley slave.

The path of development, in this as in most
other areas, was not a steady upward climb.
Even though relatively efficient mechanisms using leg muscles at good impedance matches
(figure 1.5) had been developed, sometimes
hundreds of years earlier, some designers and
manufacturers persisted in requiring heavy
hand cranking for everything from drill presses
to pneumatic diving apparatus to church-organ
blowers—even though in all these cases pedaling seems clearly advantageous.

People seem to have been thinking of human-
powered vehicles from the fifteenth century on.

A sketch attributed to a pupil of Leonardo

Figure 1.4
Medieval pump driven
by treadmill.
Reproduced, with
permission, from A. G.
Keller, *A Theatre of
Machines* (London:
Chapman and Hall,
1964).

shows a device like a bicycle fitted with pedals,
cranks, and a chain drive to the rear wheel. (As
drawn, the machine could not have been steered
and thus could not have been pedaled without
assistance in maintaining balance. It is therefore
either an inaccurate copy of an extraordinarily
brilliant and prescient Leonardo drawing or a
fraud from a much later date.) There is evidence
that a footman-propelled carriage was used in
France in the 1690s (ref. 2, p. 16). By the begin-
ning of the nineteenth century unsteerable two-
wheelers appeared in England, and these were
superseded by what is now commonly called
the hobby-horse.

It seems likely that the most important discovery in the bicycle's development was made by

Figure 1.5
Medieval bow-action
lathe, with pedal power
freeing the hands to
control turning. Courtesy
of Imperial Chemical
Industries, Ltd.

chance. Karl von Drais, who had studied mathematics and mechanics at Heidelberg but had accepted the post of master of the forests of the
Grand Duke of Baden, was intrigued by the
hobby-horses with which people were experimenting as an aid to walking the streets. He
thought that they might help him and his men
to get around the forests. Now let us speculate,
because the next crucial stage is unknown. On
streets and sidewalks, only occasionally did an
unsteerable hobby-horse have to be redirected,
by lifting the front wheel; the lack of steering
might have appeared to be a virtue. However,
for negotiating forest paths and avoiding roots,
boulders, and holes, steering must have seemed
necessary, and von Drais, whose other inventions included a binary digit system, a meat
grinder, and a typewriter, took this step (figure
1.6). Our assumption is that he had no precon-
ception that he could balance with front-wheel
steering, but simply thought that it would be a
convenience. Presumably he or one of his workers discovered the possibility of balancing one
day when going down a hill. The major discovery in bicycle history had been made, and it

was not recorded. The vehicle that von Drais developed was, however, noted in the German newspapers in 1817. It was lighter and more utilitarian than most of the heavy and somewhat ornate hobby-horses. In Paris, where von Drais obtained a five-year patent (ref. 3, p. 15), it was called the *Draisienne*. Despite some initial skepticism and ridicule, von Drais was soon demonstrating that he could exceed the speed of runners and that of the horse-pulled "posts," even over journeys of two or three hours. His ability to balance when going down inclines and to steer at speed must have been important in this. He indeed must have the principal claim to being the originator of the true bicycle.

Karl von Drais had many imitators. One was the London coachmaker Denis Johnson, whose lighter and more elegant conveyance was soon called the dandy-horse. He set up a school in which young gentlemen could learn to ride. In the next few years use of the vehicle spread to clergymen, mailmen, and tradespeople, and other mechanically minded people began taking it seriously. In 1821, Louis Gompertz fitted a swinging-arc ratchet drive to the front wheel (figure 1.7) so that the rider could pull on the steering handles to assist his feet.

Around 1839 a blacksmith named Kirkpatrick Macmillan, who lived near Dumfries, Scotland, made the first known attempt to harness leg muscles to turn the wheels directly (ref. 2, pp. 34–38). He added cranks to the rear wheels of a steerable velocipede, with connecting rods coming forward to swinging pedals (figure 1.8). Because he made it possible for the rider to pedal and stay continuously out of contact with the ground, Macmillan might be called the originator of the true bicycle. But Macmillan worked in isolation. Although he bicycled 140 miles to Glasgow on his machine (creating widespread interest, receiving the first traffic fine for knocking down a child in the throng that surrounded him, and being reported in the Glasgow papers),

Figure 1.6
A *Draisienne*. From
reference 5.

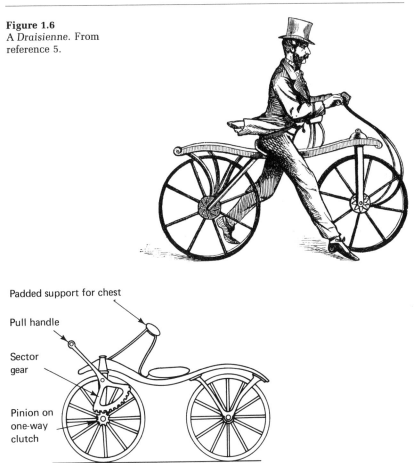

Padded support for chest

Pull handle

Sector
gear

Pinion on
one-way
clutch

Figure 1.7
Gompertz's hand drive.

Figure 1.8
A copy of Kirkpatrick Macmillan's velocipede, made around 1860 by Thomas McCall of Kilmarnock. Reproduced, with permission, from reference 2.

and although he made and sold several of his machines, no further developments followed from his efforts. The reason seems to be that the countries in which two-wheeled vehicles had been developed and received with such enthusiasm—principally Germany, France, and Britain—were in the grip of railway mania. There was a new, fast way to travel, and this technology lured the creative dreams and efforts of inventors and mechanics away from the more mundane human-powered transportation. The parallels with what was to happen sixty years later, when the enthusiasm for the safety bicycle was to evaporate before the flaming passion for the automobile, are striking.

It would be an exaggeration to claim that all development except that by Macmillan stopped. From 1815 to 1870 the term "velocipede" was used for any foot-propelled vehicle. They were used by some enthusiasts (including Prince Albert, husband of Queen Victoria), but not extensively. The machines' size and weight and the poor roads deterred walkers from changing their mode of travel. Willard Sawyer, a coachmaker in Kent, England, made increasingly sophisticated four-wheeled velocipedes, such as the one shown in figure 1.9, and exported them around the world, from about 1840 to 1870 (ref. 2, pp. 39–46). They were used by a few enthusiasts, but no movement developed. Undoubtedly there were lone mechanics and inventors in various countries making what seemed to be improvements to the Draisienne. It seems very likely that among these were some (for instance, P. M. Fischer in Schweinfurt, Germany, in 1850–1855[4]) who took what in hindsight seems the obvious step of coupling cranks and pedals to the front wheel. But all we know is that this move was left to Pierre Michaux, who also improved the rest of the machine (figure 1.10), commercialized it effectively, and set the flame that roared through France, the United States, and later Britain. The first true bicycle craze was underway.

Why, and why then? There seems to have been no major technological development to trigger it. The two-wheeled pedaled velocipede could have been invented in 1820, although the weaker metals of that time would have led to a less graceful machine. Perhaps it was helped by Michaux craftsmanship, which was widely praised. Perhaps it was Michaux's management ability; he organized factories that could produce five machines a day. Perhaps it was the Michaux family's flair for promoting the machines with demonstrations and races. But above all the machine was fun to ride, and thousands did so.

We might not think it so entrancing nowadays. The wooden wheels had rigid (compression) spokes and iron rims. It was only in the late 1860s that rubber was nailed onto the rims to cushion the harsh ride and ball bearings were first used on bicycles to give easier running. Then the French leadership was lost when, in the Franco-Prussian war of 1870–1871, the French bicycle factories were required to turn to armaments (ref. 2, p. 61).

Figure 1.9
A Sawyer four-wheeled velocipede. Reproduced, with permission, from reference 2.

Figure 1.10
A Michaux velocipede. From reference 5.

Figure 1.11
Starley's "lever-tension"
wheel. From reference 5.

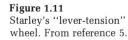

Figure 1.12
Tangent-tension spoking.
From reference 5.

Development was fast in Britain, where production had been started more to fill the unsa-
tiated French demand than to supply any
domestic market. The technical leadership was
repeatedly taken by James Starley. The suspension or tension wheel had already been experimented with in France; around 1870 Starley
introduced the "lever-tension" wheel, with radial spokes and a lever for tensioning and
torque transmission (figure 1.11), and in 1876
he came up with the logical extension of this
idea, the tangent-tension method of spoking
(figure 1.12). This has remained the standard
spoking method to this day.

Front wheels were being made larger and
larger to give a longer distance per pedal revolution, and therefore greater speed. Starley and
others recognized the advantages of using a
chain as a step-up transmission, but experimenters found that the available chains quickly
froze up in the grit and gravel of contemporary
roads. Soon front wheels were made as large as
comfortable pedaling would allow. One bought
one's bicycle to fit one's leg length. A large
"high-wheeler" or "ordinary" would have a
driving wheel about 60 inches (about 1.5 m) in

Figure 1.13
The ordinary, or high-wheeler, or penny-farthing. From reference 5.

diameter (figure 1.13). In the English-speaking world we still translate gear ratios into equivalent driving-wheel diameters, and this size corresponds to the middle gear of a typical modern bicycle. (The French use *la developpement,* the wheel's circumference.) The 1870s were the years of the dominance of the high-wheeler. By the end of the decade, ball bearings were used for both wheels and for the steering head, the rims and forks were formed from hollow tubing, the tire rubber was greatly improved over the crude type used in 1870, and the racers had been reduced to under 30 lb (13.6 kg). A ridable James "ordinary" weighing only 11 lb (5 kg) was produced.

The "ordinary" was responsible for the third two-wheeler passion, which was concentrated among the young middle-class men of France, Britain, and the United States and was fostered by military-style clubs with uniforms and even buglers. The ordinary conferred unimagined freedom on its devotees; it also engendered antipathy on the part of the majority who didn't or

couldn't bicycle. Part of the antipathy was envy.
The new freedom and style were restricted to
young men. Strict dress codes prevented all but
the most iconoclastic of women from riding
high-wheelers. Family men, even if they were
still athletic, hesitated to ride because of the fre-
quent severe injuries to riders who fell. Unath-
letic or short men were excluded automatically.
These prospective riders took to tricyles (ref. 5,
pp. 165–182), which for a time were as numer-
ous as the ordinaries.

There were two technological responses to the
need to serve the "extra-ordinary" market.
James Starley played a prominent role in the
first, and his nephew in the second.

The first was the development of practical machines of three or four wheels, in which the
need to balance was gone and the rider could be
seated in a comfortable, reasonably safe, and
perhaps more dignified position. Such vehicles
had been made at different times for at least a
century, but the old heavy construction made
propelling them a formidable task. In fact, the
motive power was often provided by one or
more servants, who in effect substituted for
horses. Starley's Coventry Lever Tricycle, patented in 1876, with his new lightweight tan-
gent-spoked wheels, could be used with
comparative ease by women in conventional
dress and by relatively staid males. Starley produced this vehicle in large numbers for several
years. In a prophetic move, he soon abandoned
lever propulsion for more conventional cranks
with circular foot motion (figure 1.14). He had
found a chain that worked, at least in the possibly more protected conditions of a tricycle. The
Coventry Lever and its successors had one large
driving wheel and two steering wheels, one in
front and one behind. Starley saw the advantage
of two large driving wheels on either side of the
rider(s) and a single steering wheel in front. For
this arrangement to work, power had to be
transmitted to two wheels, which might (for in-

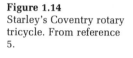

Figure 1.14
Starley's Coventry rotary
tricycle. From reference
5.

stance in a turn) be going at different speeds.
Starley reinvented the "balance gear" (ref. 5, pp.
240–241), which is now known as the differential. Starley's Royal Salvo tricycle became the
predominant form—for single riders, for two sitting side-by-side, and even for one behind the
other (figure 1.15). This is not to say that there
were no other forms; the reverse of this arrangement, for instance, with the steering wheel trailing the large driving wheels, was used for
tradesmen's carrier machines. But the front-
steerer was perceived as giving better control
(one did not have to steer toward a pedestrian
or a pothole to take avoiding action, as is necessary with rear-steerers). Gradually the front
wheel was made larger and the driving wheels
smaller, as could be done with chain drives of
increasing efficiency and reliability. By 1886 the
front wheel was connected directly to the handlebars (figure 1.16). This was a simpler, more
reliable, and safer arrangement than the rack-
and-pinion and other indirect systems which
had been used. The modern tricyle had evolved,
with the modern riding position in which one
sits or stands almost over the cranks and splits
the body weight among handlebars, pedals, and
saddle.

This modern tricyle of 1886 was also very similar to the emerging form of the modern bicycle.
In fact, the second response to the exclusion of
so many from the high-wheeler movement was

Figure 1.15
Starley's Royal Salvo
tricycle. From reference
5.

Figure 1.16
A modern-type tricycle.
From reference 5.

the development of a configuration that would make less likely a headfirst fall from a considerable height, that could be ridden in conventional dress, and that did not require gymnastic abilities.

Some improvements to the high-wheeler fulfilled only the first of these desiderata. Whatton bars (figure 1.17) were handlebars that came under the legs from behind, so that in the all-too-frequent event of a pitch forward the rider could land feet first. (Cycle clubs—but not the police—recommended that riders of high-wheelers without Whatton bars put their legs over the handlebars when going fast downhill, as in figure 1.18, for the same reason.) Some modern recumbent bicycles have similar handlebar arrangements. The designer of the American Star took the approach of making over-the-handlebars spills much less likely by putting the small wheel in front, giving it the steering function, and lowering the seating position by using a lever-and-strap drive to the large wheel through one-way clutches (figure 1.19). Unfortu-

Figure 1.17
Whatton bars. From *Cycling* (Badminton Library, 1887).

Figure 1.18
"Coasting—safe and reckless." From *Cycling* (Badminton Library, 1887).

Figure 1.19
The American Star, a
treadle-action bicycle of
1880. From L. Baudry de
Saunier, *Le cyclisme,
théorique et pratique*
(Paris: Librairie Illustré,
1892).

Figure 1.19
The American Star, a
treadle-action bicycle of
1880. From L. Baudry de
Saunier, *Le cyclisme,
théorique et pratique*
(Paris: Librairie Illustré,
1892).

nately, this arrived too late (1885) to have much impact, because the true "safety" bicycle had evolved almost to its modern form by that date. Another type of bicycle that was safer to ride than the high ordinary was the "dwarf" front-driver (figure 1.20) with a geared-up drive to a smaller front wheel (ref. 5, pp. 152, 158). Such "geared ordinaries" were offered in the early 1890s because riders accustomed to front-drive machines did not always take kindly to the rear-drive safeties. Small-wheeled Bantam bicycles with an epicyclic hub gear (figure 1.21) were marketed as late as 1900. These can be classed as the earliest "portable" machines, because they fitted well into the "boot" of a horse-drawn "trap."

It had long been recognized that it would be most desirable from the viewpoint of safety to have the rider sitting between two wheels of moderate size. Many attempts were made over the years. Macmillan's lever-propelled velocipede of 1840 had this configuration. In 1869—the year of the first Paris velocipede show, at which rubber tires, variable gears, freewheels, tubular frames, sprung wheels, and band brakes were shown—Andre Guilment made what might

Figure 1.20
"Dwarf" front-drive
bicycle. From reference
5.

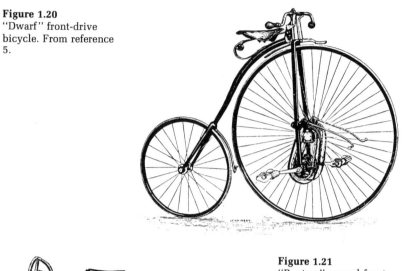

Figure 1.21
"Bantam" geared front-
drive safety bicycle.
From reference 5.

Figure 1.22
Starley safety bicycle.
From reference 5.

be classed as the first safety bicycle. But the direct descendants of today's bicycles evolved rapidly in the one or two years before 1885, when several were shown in Britain's annual Stanley Bicycle Show. James Starley had died in 1881, but his nephew John Kemp Starley, working with William Sutton, produced a series of "Rover" safety bicycles, which by 1886 had direct steering and something very close to the diamond frame used in most bicycles today (figure 1.22).

One major development in the mainstream flowing to the modern bicycle remained: the pneumatic tire. This was patented in 1888 by John Boyd Dunlop, a Scottish veterinarian in Belfast, although another Scot, R. W. Thomson, had patented, but apparently not developed, pneumatic tires for horse-drawn vehicles in 1845.[6] Dunlop's early tires (made to smooth the ride of his son's tricycle) were crude, but by May 1889 they were used by W. Hume in bicycle races in Belfast—and he won four out of four. Success in racing in those days gave a clear signal to a public confused by a multitude of diverse developments. Bicyclists saw that, as in the case of the "safety" versus the high-wheeled bicycle, a development had arrived that promised not only greater speed, or the same speed with less effort, but greater comfort and, especially, greater safety. Within four years, solid tires had virtually disappeared from new bicycles, and Dunlop was a sterling millionaire.

With the arrival of the pneumatic-tired direct-steering safety bicycle, only refinements in components remained to be accomplished before the modern-day bicycle could be said to have been fully developed. Various types of epicyclic spur-gear variable-ratio transmissions for the brackets and rear hubs of chain-driven safety bicycles came on the market in Britain in the 1890s. Some heavier devices were available earlier for tricycles. The Sturmey-Archer three-speed hub

(1902) was the predominant type, as it still is in many parts of the world, but there were many competitors at around the turn of the century. The derailleur or shifting-chain gear appeared in one form at about this time but was not popular. It was developed by degrees in Europe, and was eventually accepted for racing in the 1920s.

Undoubtedly, much more will be discovered about the history of the modern traditional single-rider bicycle, and unrecognized inventors will receive the honor due them. Inquiring readers can find much more history than we have space for here in the excellent books listed at the end of the chapter.

We close this chapter with a short review of the history of one of the many types of nontraditional bicycles: the "recumbent." Our reason for discussing recumbents rather than tandems, folding bicycles, pedicabs, goods transporters, or sprung bicycles is that most modern record-breaking machines are recumbents. Also, D.G.W. is convinced that greater safety can result from the use of the recumbent riding position in highway bicycles. In addition, what little we know of the history of this variant form might help to illustrate the past and present flavor of the cycle industry.

Many early cycles (particular tricycles) used the semirecumbent position. The "boneshaker" was often ridden with the saddle well back on the backbone spring and the feet at an angle considerably higher than that for the modern upright "safety." In contrast with the riders of the high-wheeler and of the "safety," who were told to position the center of gravity vertically over the center of the crank, the semirecumbent rider sits in something like a chair and puts his feet out forward on the pedals. The pedal-force reaction is taken not by the weight of the body (or, when that is exceeded, by pulling down on the handlebars), but by the backrest.

The first known semirecumbent bicycle (by

which we mean one where the rider's center of gravity was low enough relative to the front-wheel road-contact point for there to be a negligibly low possibility of his being thrown over the front wheel in an accident) was built in Ghent by Challand sometime before 1895 (ref. 3, p. 47). Challand called it the *Normal Bicyclette*. The rider sat rather high, directly over the rear wheel. In 1896 a U.S. patent application was filed by I. F. Wales for a somewhat strange-looking recumbent bicycle with hand and foot drive (figure 1.23).[7] A much more modern-looking recumbent bicycle was constructed by an American named Brown and taken to Britain in 1901 (figure 1.24).[8] By this time orthodoxy rested firmly with the traditional safety bicycle, and the derision that had successively greeted the hobby-horse, the *Draisienne*, the velocipede, and the safety had been forgotten. A review of the Brown recumbent in *The Cyclist* (ref. 8) was derisive to the point of sarcasm:

> . . . the curiously unsuitable monstrosity in the way of a novel bicycle shown in the single existing example of Mr. Brown's idea of the cycle of the future here illustrated. . . . The illustration(s) fully show(s) the rider's position and the general construction of this crazy effort. . . . The weight (30 lb) and cost of the machine are greatly increased. . . . The mounting and dismounting are easy, and this is a fine coasting machine, the great wheelbase making very smooth riding . . . and turns in a small circle. The machine runs light and is a good hill-climber, and it is only fair to say that the general action of this queerest of all attempts at cycle improvement is easy and good—far better than its appearance indicates. . . . The surprising fact is that any man in his sober senses could believe that there was a market for this long and heavy monstrosity at the price of a hundred dollars (£20). . . .

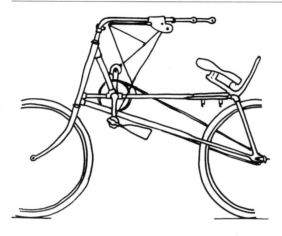

Figure 1.23
Design for hand-and-foot-powered recumbent bicycle patented by I. F. Wales in 1897.

Figure 1.24
Brown's 1900 recumbent bicycle. Adapted from reference 8.

What of the apparent lack of American contributions to the mainstream of bicycle development? What happened to the Yankee genius in engineering and mechanics? The U.S. patent office was in fact flooded with applications to patent improvements to velocipedes from 1868 on. The French and British makers found it necessary to follow the developments taking place across the Atlantic (ref. 3, p. 61 et seq.). In 1869 Pickering's Improved Velocipedes were exported from New York to Liverpool. But the American craze, which the *Scientific American* stated had made the art of walking obsolete, suddenly petered out in 1871 as quickly as it had started, leaving new businesses bankrupt and inventors with nowhere to go (ref. 2, p. 66). There was then a lull until 1877, when the high-wheel bicycle was imported. Colonel Albert Pope started manufacturing them in the East a year later. But conditions were difficult for bicycles. In Europe, the high bicycle enabled people to travel much farther than was comfortably possible on a velocipede, and in Britain the roads were good enough for the country to be traversed from Lands End in southwest Cornwall to John O'Groats in northeast Scotland (924 miles; 1,490 km) in seven days (ref. 2, pp. 126–127). In the United States the distances between towns were (except perhaps in New England) enormous, and the roads were poor (ref. 2, pp. 82–83). Accordingly, the bicycle did not have, and did not convey, as much freedom, and the market was therefore smaller and far more dispersed than in Europe.

Recumbents were more successful in Europe. Peugeot produced one model commercially in 1914, but this effort was doubtless snuffed out by the start that year of the Great War. After the war, the Swiss Zeppelin engineer Paul Jaray built recumbents in Stuttgart in 1921.[9]

Racing recumbents (figure 1.25) were brought out in France in the 1930s. They became known as "velocars," probably because four-wheeled

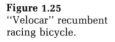

Figure 1.25
"Velocar" recumbent
racing bicycle.

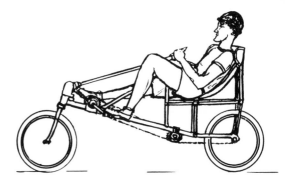

vehicles of that name had become popular, and
these used a similar position for the rider. With
a velocar, a relatively unknown racing cyclist,
Francis Faure, defeated the world champion,
Lemoire, in a 4-km pursuit race and broke track
records that had been established on conventional machines.[10] A genuine orthodoxy per-
vaded the bicycle industry and the International
Cycling Union, which controlled world bicycle
racing. Instead of setting up a procedure and
special category for machines such as the velocar, the Union banned unconventional types
from organized competition. This decision de-
nied novel ideas the opportunity of being tested
and publicized through racing, and thereby deterred experimentation and development.

Only with the open-rule human-powered-vehicle competitions, started in California in 1974,
has the inventiveness of human-powered-vehicle designers been given an incentive. With all
classes of races now being won by recumbent
machines of a large variety of types, the technological history of this type of vehicle, and of bicycles in general, is again being written. These
are exciting times. We wonder if there may not
also be a parallel in this new period of development with the period that started around 1866.
The excitement over railway travel had seemed
to drain away either the excess energies of inventors or the support for their activities, so that

bicycle development languished. Occasional inventions like Gompertz's or Macmillan's were not followed up. But perhaps by the mid-1860s the railway was accepted, and it was apparent that it was not going to solve all transportation problems. Similarly, in the 1890s the motorcar arrived, and suddenly it was fashionable not only to travel in them, but to be involved in developing them. And two bicycle mechanics produced the first powered airplane only a little later. From then almost until the present day there has been a widely acknowledged love affair with the automobile, and with the airplane, first in the developed countries and later in the undeveloped countries. Only when disenchantment set in over the damage which these methods of transportation were wreaking on our cities did widespread enthusiasm for bicycle development surface once more.

May future histories record that new developments led to a new wave of popularity for human-powered travel, one that will last longer than some of the crazes of the past.

References

1. L. Reti (ed.), *The Unknown Leonardo* (New York: McGraw-Hill, 1974), pp. 178–179.

2. A. Ritchie, *King of the Road* (Berkeley, Calif.: Ten-Speed, 1975).

3. W. Wolf, *Fahrrad und Radfahrer* (Leipzig: Spamer, 1890/Dortmund: Hitzegrad, 1979).

4. P. von Salvisberg, *Der Radfahrsport in Bild und Wort* (Munich, 1897/Hildesheim and New York: Olms, 1980), p. 13.

5. A. Sharp, *Bicycles and Tricycles* (London: Longmans, Green, 1896/Cambridge, Mass.: MIT Press, 1977).

6. R. W. Thomson, Carriage Wheels, U.K. patent 10,990, 1845.

7. R. Barrett, Recumbent cycles, *The Boneshaker* (Southern Veteran-Cycle Club, U.K.) 7 (1972): 227–243.

8. H. Dolnar, An American stroke for novelty, *The Cyclist* (London) (8 January 1902): 20.

9. H.E. Lessing (University of Ulm, West Germany), personal letter to D.G.W., 14 August 1980.

10. "The Loiterer," Velocar versus normal, *Cycling* (London) (2 March 1934): 202.

Recommended reading   S. S. Wilson, Bicycle technology, *Scientific American* (March 1973): 81–92.

G. H. Bowden, *The Story of the Raleigh Cycle* (London: Allen, 1975).

———. *A Shortened History of the Bicycle* (Nottingham: Raleigh).

I. A. Leonard, *When Bikehood Was in Flower* (South Tamworth, N.H.: Bearcamp, 1969).

2　Human power generation

As an energy producer, the human body has similarities and dissimilarities with the engine of an automobile. Energy is taken in through fuel (food and drink, in the case of humans). "Useful" energy is put out in the form of torque on a crankshaft; and "waste" energy is dissipated as heat, which may be beneficial in cold weather. The peak efficiencies of the two systems (the energy in the power going to the crankshaft divided by the energy in the food or fuel) are remarkably close to one another, in the region of 20–30 percent. But automobile engines seldom work at peak efficiency, and only at full power, whereas the rider of a multispeed bicycle can operate much closer to peak efficiency at all times. Another significant difference is that whereas the automobile is powered by a "heat engine," the human body is a kind of fuel cell. Also, human output changes over time, and can draw on body reserves; the gasoline engine can work steadily until the fuel runs out, when the engine delivers nothing. Humans also vary greatly from one to another, and from one day to another, and from one life stage to another.

Most of the information in this chapter has been obtained by careful experiments, most often with test subjects on power-output-measuring devices called ergometers (figures 2.1, 2.2). Most ergometers are pedaled in the same way as bicycles; other types are "rowed" or "walked." Exercise physiologists can take careful and often precise measurements of human work output in their laboratories. However, we must keep in mind three reservations about ergometers:

- People vary widely in performance, and unless very many are tested (as has been seldom the

Figure 2.1
Racing-bicycle ergometer.

case) the data cannot be generalized to the whole of humanity.

- Pedaling or rowing an ergometer usually feels stranger than riding a novel type of bicycle. It may take a month of regular riding before one becomes accustomed to and efficient with a novel bicycle, as one's muscle actions gradually adapt to a new motion. Subjects are seldom given the opportunity to adapt for more than a few minutes (occasionally, hours) to working an ergometer.

- Most of the energy put into bicycling, and a fair proportion of that put into rowing a boat, goes into air friction, and the heat transferred from a hot body to a cooler airstream is largely proportional to air friction. Subjects pedaling ergometers are seldom given equivalent cooling, and their maximum output is therefore likely to be limited by heat stress. (There are exercisers on the U.S. market in which most of the power is dissipated in fans,

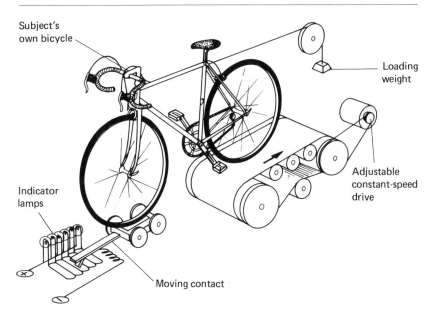

Subject's
own bicycle

Loading
weight

Adjustable
constant-speed
drive

Indicator
lamps

Moving contact

Figure 2.2
Müller ergometer. Load
and speed are set; subject
tries to keep center lamp
lit. Run stops when
rearmost lamp lights up.

thus simulating the "cube-law" effect of wind
resistance, but the air flow is not directed at
the pedaler).

For these reasons, power-output measurements
on ergometers are likely to be lower than would
be given by the same subjects pedaling or rowing their own familiar machines in a reasonably
cool breeze.

Some of the available test data on human
power output are, however, taken from subjects
bicycling on pavement, with various ingenious
means used to measure work output and/or oxygen consumption (figure 2.3). These data are
likely to be more reliable than ergometer data.
Even here someone wearing various sensors,
possibly including a breathing mask, is likely to
find that at maximum output there is some degree of resistance to movement and/or to breathing, and that this will reduce the peak output
somewhat.[1]

Figure 2.3
Cyclist using breathing-
rate-measuring
equipment. Photograph
supplied by research
workers at Nijmegen
University, The
Netherlands.

Most ergometers have frames, saddles, handlebars, and cranks similar to those of ordinary bicycles. The crank drives some form of resistance
or brake, and the whole device is fastened to a
stand, which remains stationary during use.
Other ergometers can measure the output from
hand-cranking in addition to that from pedaling. Some permit various types of foot motion
and body reaction, including rowing (sliding-
seat) actions. The methods employed for power
measurement range from the crude to the sophisticated. One problem is that human leg-
power output varies cyclically (as does that of a
piston engine) rather than being smooth (as

with a turbine). A device indicating instantaneous power (pedal force in the direction of motion multiplied by pedal velocity) would show peak values of perhaps one horsepower (746 watts), whereas the average would be only 0.2 hp (149.2 W). Therefore, some form of averaging is usually employed. In some cases the subject is supposed to keep pedaling at a constant rate over a minute or two to obtain accurate results; in other systems the power can be integrated and averaged electronically over any desired number of crank revolutions.[2,3]

Muscle chemistry and mechanisms

A large muscle is composed of a large number of individual fibers. Each fiber, like the muscle itself, can only contract; a muscle cannot "push." Neither can an individual muscle fiber exert a continuous force. It is caused to contract by the nervous system's release of adenosine triphosphate (ATP). After contraction, a fiber will again relax. If a muscle is required to exert a continuous force, for instance in holding up a weight, muscle fibers will "fire" sequentially. Even if the weight is not lifted, which means that in the thermodynamic sense no external work is being done, the muscle will require energy either from its stores or from the bloodstream. We call this "isometric" exercise, because there is no change in the measurement of the muscle or of the body. If we are trying to maximize external work and to minimize fatigue, as we are in bicycling, we should avoid isometric stress as far as possible.

The ATP, which is the muscle fiber's immediate fuel, can be supplied in two ways.[4] For almost immediate short-term use, the muscle can draw on its own stored phosphoryl creatine and glycogen. It can use these without the need for oxygen from the blood; hence, we call this muscle action *anaerobic*. The muscle fibers that work anaerobically are termed type-II fibers. They are developed by sprinters and by animals who rely on a sudden spurt of activity to escape from their predators. These fibers are found in the white meat of the turkey. Anaerobic-muscle use in humans can last for up to about 5 minutes. Because there is a restricted amount of energy available (proportional to the mass of the muscle), the duration of its use depends on the power output demanded. For longer-term use, in so-called steady state, the ATP needed by the muscle must be supplied from glucose and fatty acids that are supplied by the blood and oxidized. The muscle fibers that can work for long periods, which use the blood and work with oxygen and therefore work *aerobically,* are termed type-I fibers, and are dark brown, like the dark meat of a turkey's legs. Cyclists need both type-I and type-II fibers, and can develop one type more than the other by training and adaptation.

Breathing

When a cyclist is tested on an ergometer for a sufficiently long period that the aerobic muscle actions predominate, it is found (see, for example, ref. 5) that for each milliliter per second of oxygen absorbed by the lungs about 4.5 watts of power are put out by the legs. Laboratory experiments on the calorific ("heating") value of the blood sugars and other chemicals oxidized by the absorbed oxygen show that for the same flow the heat output would be about 18 W. The efficiency of muscle action is, therefore, roughly 25 percent.

Most of the 75 percent of the energy that does not appear as power at the pedals is dissipated as heat. The human body employs various mechanisms for keeping the trunk warm in cold weather when no exercise is being carried out, and other mechanisms for limiting the temperature rise to about 2°C in hot weather and during heavy physical activity.[6-11] The evaporation of perspiration can dissipate an enormous amount of heat: 2.42 kilowatts per gram per second of perspiration. Thus, it is important for exercising humans that their perspiration evaporate and not just drip off. Fast-moving air evaporates water far more quickly than slow-moving air. As a consequence, a pedaler on a stationary ergometer drips sweat profusely at a work rate of 0.5 hp (373 W). At 27 mph (12.1 m/sec)—a speed corresponding to 373 W—a riding bicyclist is cooled far more effectively by sweat evaporation (refs. 2, 12).

Thermodynamic engines such as steam-turbine plants and internal-combustion engines are also usually only about 20–30 percent efficient in converting fuel energy to mechanical work, although the best engines working in optimum conditions can attain an efficiency of 45 percent. However, the limitations here derive from the second law of thermodynamics, and therefore from the levels at which heat is added to and rejected from the engine.

One of the many ways of expressing the second law of thermodynamics is the following: No engine can be more efficient than a thermodynamically reversible engine, and the efficiency of such an engine can be shown to be given by

$$\frac{\text{Power output } W}{\text{Rate of heat input } Q_2}$$
$$= \frac{Q_2 - Q_1}{Q_2} \text{ for all engines}$$
$$= \frac{T_2 - T_1}{T_2} \text{ for reversible perfect engines,}$$

where T_2 is the temperature and Q_2 the rate of heat addition, and T_1 is the temperature and Q_1 the rate of heat rejection. Temperatures are given in degrees above absolute zero (degrees Rankine, °R, on the Fahrenheit scale, degrees Kelvin, °K, on the Celsius scale). Absolute zero is $-460°F$ ($-273°C$).

A steam-turbine plant fed with high-temperature steam is more efficient than one using steam at lower temperatures. To achieve a thermodynamic efficiency of 25 percent, even an ideal engine rejecting heat at above room temperature T_1 (as must the human body) would require that its fuel energy be absorbed at T_2, which can be calculated as follows:

$$\text{Efficiency} \equiv 1 - \frac{T_1}{T_2} = 0.25 = \tfrac{1}{4};$$
$$T_2 = \frac{4T_1}{3} = \frac{4 \times 300°\text{K}}{3}$$
$$= 400°\text{K} \ (127°\text{C}; 720°\text{R}; 260°\text{F})$$

for a heat-rejection temperature T_1 of 27°C (300°K; 80°F).

Obviously, 127°C cannot be tolerated in the body. Therefore, the human "engine" is not subject to the restrictions of the second law of thermodynamics. It is a type of fuel cell in which chemical energy is converted directly to mechanical power. The energy not converted to power must appear, as for heat engines and fuel cells, as heat.

The human engine has an additional characteristic not generally found in machines: Some fuel must be "burned" to keep it going when it is at rest. (In this sense it is somewhat similar to a traditional steam plant, in which fuel must be burned continually to keep steam pressure up even when no power is being delivered). The amount of oxygen absorbed by the lungs of a person of average weight, at rest and not using any voluntary muscles, is about 5.5 milliliters per second (one-third of a liter per minute). This quantity is additional to any other absorption from muscle exercise. In ordinary air, a liter of oxygen is found in about 5 liters of air. However, when air is breathed, about 24 liters must be passed through the lungs for a liter of oxygen to be absorbed.[13] Thus, about 380 percent more air than is needed to produce energy is used in the human engine. Most other engines, such as internal-combustion and steam engines, require only about 5–10 percent "excess" air to ensure complete combustion of the fuel. Gas turbines more nearly approach human lungs, taking in about 200 percent excess air.

Table 2.1 Breathing rates for cycling and walking.

Cycling Speed				Tractive power		Breathing rate (l/min)		Metabolic heat	
Racer[a]		Tourist[b]							
mph	m/sec	mph	m/sec	hp	W	Oxygen	Air	kcal/min	W
27	12.1	22.5	10.1	0.5	373	4.8	115	24	1,680
25	11.2	21	9.4	0.4	298	3.4	93	19.5	1,365
22	9.8	18.5	8.3	0.3	224	3	72	15	1,050
19	8.5	16	7.2	0.2	149	2.1	50	10.5	735
14.5	6.5	12	5.4	0.11	82	1.2	29	6	420
10.5	4.7	8.3	3.7	0.05	37	0.75	18	3.75	263
7.2	3.2	6	2.7	0.025	19	0.53	13	2.65	186
3.2	1.4	1.8	0.8	0.008	6	0.38	9	1.9	133
0	0	0	0	0	0	0.3	7	1.5	105
Walking Speed									
mph	m/sec								
4.46	2			0.141	105	1.83	44	9.1	637
3.33	1.5			0.076	57	1.1	26	5.5	385
2.23	1			0.0415	31	0.71	18	3.5	245
1.1	0.5			0.0226	17	0.52	12.5	2.5	175
0	0			0	0	0.28	6.8	1.4	98

Sources of data "Velox," *Velocipedes. Bicycles and Tricycles: How to Make and Use Them* (London: Routledge, 1869); M. G. Bekker, *Theory of Land Locomotion* (Ann Arbor: University of Michigan Press, 1962); G. A. Dean, An analysis of the energy expenditure in level and grade walking, *Ergonomics* 8 (1965), no. 1: 31–47.

a. Total mass 77 kg (170 lb), frontal area 0.34 m^2 (3.6 ft^2), tire pressure 100 lbf/m^2 (689 kPa).
b. Total mass 85 kg (187 lb), frontal area 0.511 m^2 (5.5 ft^2), tire pressure 50 lbf/m^2 (345 kPa).

Using all the above information, we show in
table 2.1 how breathing rates increase for an average rider (150 lb; 68.04 kg) cycling on the
level in still air. It is assumed that, for every
liter of oxygen absorbed, 24 liters of air have to
be breathed.

For a nonathletic person the maximum oxygen-breathing rate is assumed to be about 50
ml/sec, or 3 l/min. Table 2.1 shows that when a
rider is using about half the maximum oxygen-
breathing capacity the power output is about
0.1 hp (74.6 W). These conditions are thought to
be such that an average fit man or woman could
work for several hours without suffering fatigue
to an extent from which reasonably rapid recovery is not possible. This rate of work is recommended for workers in mines (refs. 5, 14).
Experience has also shown that 0.1 hp (74.6 W)
propels a rider at about 12 mph (5.36 m/sec) on
a lightweight touring bicycle. As this speed can
ordinarily be maintained by experienced but average touring-type riders, the numbers given in
table 2.1 seem sound. Miscellaneous data given
by Adams (ref. 14) and Harrison (ref. 15) show
average heat loads of 290–630 W for speeds of
$6\frac{1}{2}$–13 mph (2.9–5.8 m/sec); some of these and
other data are collected in figure 2.4

Breathing effectiveness decreases with age. An
athlete's peak is reached at about 20, and it is a
rule of thumb that breathing capacity is halved
by age 80.[16] This figure has been substantiated,
and the shape of the capacity-reduction curve
has been established, through analysis of the
U.K. 1971 50-mile amateur time trials, in which
the ages of the best "all-rounders" and of the
"veterans" were given. The average speed for
each rider is plotted against the rider's age in
figure 2.5. There is no recognizable falloff in
performance up to age 40, after which there is a
steady drop to that for the oldest competitor,
aged 77. These performances have been converted to breathing capacity, estimated by the
method of reference 6. When the curve is ex-

Figure 2.4
Gross caloric expenditures of bicyclists. Data for points from L. Zuntz, *Untersuchungen über den Gasswechsel und Engegesumsatz des Radfahrer* (Berlin: Hirschwald, 1899); D. B. Dill, J. C. Seed, and Z. N. Marzulli, Energy expenditure in bicycle riding, *J. Appl. Physiol.* 7 (1954): 320–324; O. G. Edholm, J. G. Fletcher, E.

M. Widdowson, and R. A. MacCanee, Energy expenditure and food intake of individual men, *Br. J. Nutrition* 9 (1955): 286–300; M. S. Malhotra, S. S. Ramaswany, and S. N. Ray, Influence of body weight on energy expenditure, *J. Appl. Physiol.* 12 (1962): 193–235; J. D. Brooke and C. J. Davies, Comment on "The estimation of energy expenditure of sporting cyclists,"

Ergonomics 16 (1973), no. 2: 237–238; and reference 14. Curves A and B (estimations for tractive-resistance calculations) from reference 6. Curve C based on data from G. A. Dean, An analysis of the energy expenditure in level and grade walking, *Ergonomics* 8 (1965), no. 1: 31–47, and J. S. Haldane, *Respiration* (London: Oxford University Press, 1922).

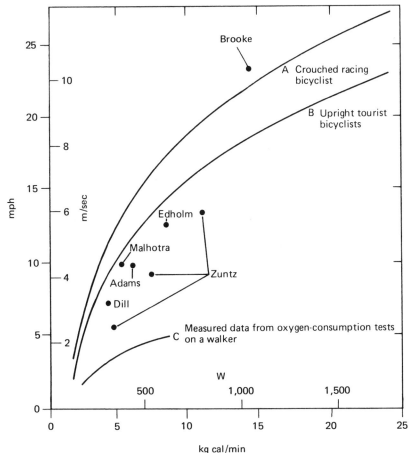

Figure 2.5
Average speeds and
estimated breathing
capacities in 50-mile
trials (1971) as a function
of age.

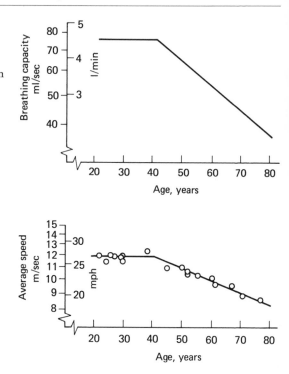

trapolated to 80 years, the estimated breathing
capacity is indeed very close to half the peak
value. (These results are for athletes. It is easy,
through disease, smoking, or lack of exercise, to
diminish one's breathing capacity by a much
greater degree than that shown.)

These data are also given some confirmation
by the performances at different times of Reg
Harris, the former world champion sprint bicyclist. In his peak form from 25 to 35 years of
age he would reach about 40 mph (17.9 m/sec)
for the final 200-meter sprint on the track. (In
this final spurt his muscles would largely be
working anaerobically). At age 55 he could
achieve about 36 mph (16.1 m/sec) (ref. 7). This
10-percent reduction in speed agrees closely
with figure 2.3, and is equivalent to a reduction
in power requirement of about 25 percent. Such

a reduction in maximum power-output capacity for the same span of years is predicted by Falls using Müller's data (ref. 7, p. 304).

Road-racing cyclists appear to be able to use about 80 percent of maximum breathing capacity for several hours.

Up to a breathing rate of about 0.67 ml/sec (40 l/min), people tend to breathe through the nose (ref. 7, p. 55) if they have healthy nasal passages. Nasal passages usually open during exercise, even during a heavy cold. Above this rate, the resistance to flow of even a healthy nose becomes penalizing, and mouth breathing is substituted. For a normally healthy individual riding on the level in still air on a lightweight bicycle, this limiting rate for nasal breathing is reached at about 14 mph (6.3 m/sec).

Tests by Pugh[17] on bicyclists riding on an ergometer and on a flat concrete track at speeds up to 27 mph (12.1 m/sec) confirmed the data of table 2.1. Pugh's work also confirmed 23.6 percent as the net muscle efficiency for oxygen use (ref. 6). This figure was used in table 2.1 to calculate the metabolic heat rates expended by the rider from the tractive forces at the driving wheel. The net efficiency therefore includes the transmission losses from the rider's foot to the contact point of the rear tire with the road—losses at the pedal, the crank set, the chain, and the wheel hub. If one includes in the wheel-bearing losses those due to the load reaction as well as to chain-force reaction, the total loss is about 5 percent of the rider's output for a bicycle in first-class condition.

Maximum performance versus time

The power output of any animal will start at a maximum as muscles draw on anaerobic reserves, and will fall to the steady-state, aerobic level. Even aerobically, we would expect a falling work output with time because of fatigue (which appears to be due to a clogging of the muscle "drainage"—the lymphatic system—with the breakdown products of the ATP, principally lactic acid). Some of the best ergometer data (figure 2.6) were taken by Harrison (ref. 15) with nine fit men, not champion cyclists or oarsmen, aged 22–42. We judge these data to be good because the highest outputs—apparently those of Harrison himself—tend to form an envelope around the data of others. We noted above that there are many reasons why the power outputs of people as measured on ergometers might be less than the peaks of which they would be capable on a bicycle or in a rowing shell; Harrison, who designed an ingenious ergometer capable of many different foot and hand motions and used a conventional bicycling ergometer, must have avoided the pitfalls.

Harrison's curve for normal pedaling or cycling (figure 2.6, curve 1) agrees closely with Nonweiler's[18] estimated curve for racing cyclists (figure 2.7, curve A). The curves for linear ("rowing") foot motion (2 and 4 in figure 2.6) are initially considerably below the cycling curve but approach it after 5 minutes. Rowing data taken on an ergometer have an additional reason for a diminished output: If the feet are fixed with respect to the ground, as they are normally fixed to the boat, the rower must accelerate his body and then use his muscle energy to reverse the acceleration—a wasteful process. This occurs to only a minor extent in actual rowing. A rowing shell is so light that the center of gravity of the body is little displaced, and the boat is accelerated and decelerated quite strongly. This wastes some energy, but not nearly so much as in a stationary ergometer. A bicycle propelled by a rowing motion would also have a highly variable velocity. The variability would be more pronounced if the feet were fixed to the bicycle and the seat were on a roller track (the usual arrangement in a shell) than if the seat were fixed and the feet were on a track. Likewise, in an ergometer one would expect the power output to be less when the feet were fixed to the stationary frame (as

Figure 2.6
Human power by various motions: cycling (curve 1), free and forced rowing with feet fixed (curves 2 and 3, respectively), and free and forced rowing with seat fixed (curves 4 and 5, respectively). From reference 15.

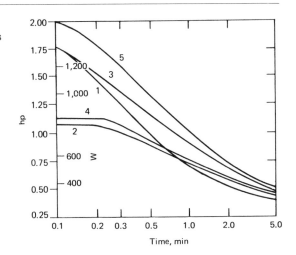

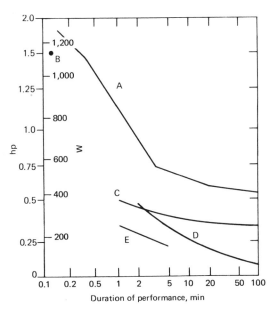

Figure 2.7
Peak human power output for different durations. Curve A: estimated cycling data from reference 18. Point B: ergometer data from Loughborough University (pers. comm.). Curve C: ergometer data from T. Nonweiler, Air Resistance of Racing Cyclists, report 106, College of Aeronautics, Cranfield, England, 1956. Curve D: winch data from J. C. Trautwine, *The Civil Engineer's Reference book*, 21st edition (Ithaca, N.Y.: Trautwine, 1937), pp. 685–687. Curve E: ergometer hand-crank data from reference 12.

Harrison found; see reference 15) than if the seat were fixed and feet moved.

Of great interest are Harrison's results for what he called "forced" rowing. (This has nothing to do with the slave galleys mentioned in chapter 1.) Harrison set up a motion whereby the mechanism defined the ends of the stroke and conserved the kinetic energy of the moving masses. The piston-crank mechanism of a car engine is of this type. With forced rowing and the seat fixed, about 12.5 percent more power than with normal pedaling was obtained throughout the time period for all subjects. This significant finding has not yet been translated into a practical mechanism for harnessing pedal power, despite several attempts by D.G.W.

Nonweiler, besides estimating power output during cycling (figure 2.7, curve A), obtained ergometer data (curve C), which were considerably lower in output. The reasons for the decrement may include those listed earlier as factors that might reduce ergometer output. For reference, figure 2.7 also gives curves for winching and for hand cranking.

The question frequently arises as to whether or not one can add hand cranking to pedaling and obtain a total power output equal to what one would produce using each mode independently. Kyle and co-workers showed that, for periods of up to a minute, 11–18 percent more power than with the legs alone could be obtained with hand and foot cranking.[19] The power was greater when the arms and legs were cranking out of phase than when each arm moved together with the leg on that side. Whether or not this gain can be projected beyond the period of anaerobic work is not known.

Bicycling performance   Most ergometer tests are made with subjects who are young, male, and near the championship class. One reason is obvious: A performance lower than that given by champions might be due to lesser ability, or to any of the deficiencies in the testing method detailed above. (Harrison's data are remarkable in recording high performance by nonathletes.) We report here two studies that appear to have been carefully made, used nonathletes, and investigated various parameters such as the effect of pedaling rate to the extent that interesting comparisons with the performance of athletes can be drawn.

Effect of pedaling rate

Grosse-Lordemann and Müller[20] conducted ergometer tests using the subjects' own bicycles, as in figure 2.2. This method ruled out unfamiliarity with the foot motions and riding positions. The output was measured at the wheels, and therefore was affected by transmission and tire-rolling losses. Figure 2.8 shows the power-time curves for a 34-year-old man, and also the pedal rotation rates (which could be preset on the ergometer). The subject developed maximum power for all durations at 40–50 rpm, a

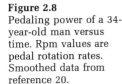

Figure 2.8
Pedaling power of a 34-year-old man versus time. Rpm values are pedal rotation rates. Smoothed data from reference 20.

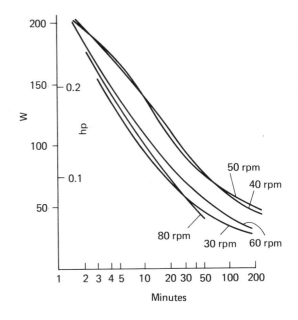

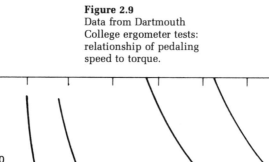

Figure 2.9
Data from Dartmouth
College ergometer tests:
relationship of pedaling
speed to torque.

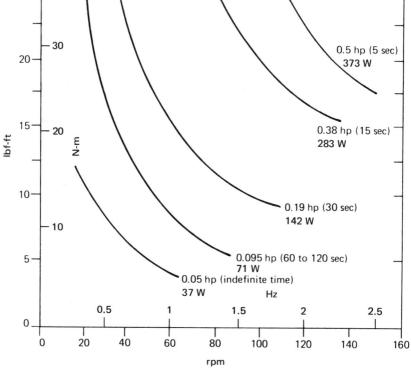

rotation rate considerably slower than those
found for peak-output short-duration pedaling
by other investigators. The power outputs were
also well below those found by others. In these
tests, no attempt was made to test champion-
ship-class riders. Garry and Wishart[21] also
found that maximum muscle efficiency was
achieved at about 50 rpm.

Students at Dartmouth College used an ergometer to find out what power output an ordinary untrained bicyclist could maintain over
useful periods of time.[22] They found that for
prolonged periods about 0.05 hp (37.3 W) was
maintained with pedaling rates of 20–60 rpm
(figure 2.9). It can be calculated that this power
would give a road speed, on the level with no
wind, of about 8 mph (3.6 m/sec). This speed is
commonly achieved by an average "utility" bicyclist and therefore provides a check on the
power measurement. This power result and
other powers tolerable to the Dartmouth bicyclists for briefer periods are shown as the nearly
straight lines in figure 2.10. The expenditure of
0.05 hp (37.3 W) can be achieved over a range
of pedaling speeds from about 30 to 60 rpm.
Therefore, as experience has shown, precise
gear selection is not necessary for utility bicycling. Japanese experimental data[23] confirming
this finding are plotted in figure 2.11.

Other ergometer experiments similar to the
Dartmouth tests were conducted by Wilkie;
these are summarized on page 8 of reference 24.
Wilkie's subjects were instructed to exert themselves to the limit in order to record their maximum power outputs for varying periods of time.
The peak power obtained was 0.54 hp (402.7 W)
for one minute, and for 60–270 minutes the
powers were 0.08–0.19 hp (59.7–141.7 W).
These powers are somewhat above those of the
Dartmouth students and close to those recorded
for laborers turning winches (figure 2.8). It appears logical to take the Dartmouth results as

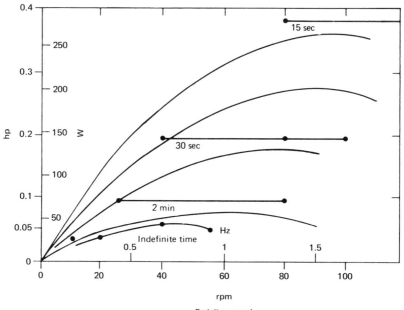

rpm

Pedaling speed

Figure 2.10
Data from Dartmouth
College ergometer tests:
power output as a
function of pedaling
speed. Horizontal straight
lines show maximum
power of an average
pedaler for the duration
noted; curves (except that
for indefinite time) are
based on data from W.
Brown, Cycle gearing in
theory and practice,
Cycling (5 July 1944):
12–13.

more indicative of the power output of an average untrained "utility" rider.

Workers at Nijmegen University measured the breathing rates of riders during actual bicycling, as in figure 2.3, and compared the results with the data of Hermans-Telvy and Binkhorst[25] on walking and running (figure 2.12). The gears used with a standard three-speed hub are indicated for the bicycling results. It is clear, for instance, from the group of three first-gear points at about 14 mph (6.3 m/sec) that the use of a low gear at these speeds is less efficient. The pedaling rate would be about 98 rpm, as against 66 in second gear and 56 in top gear. In all cases the power output would be about 134 W (0.18 hp), in agreement with the data at this power in figure 2.11. Thus, the common advice to riders to "keep spinning" (to pedal at a high rate) is appropriate only for maximum-speed, maximum-output sprinting. This is confirmed by results of the once-popular 25-mile time trials in Britain, in which riders were restricted to 70-inch (1.78-m) gears with no free-wheel. The speeds achieved were a few percent lower than those achieved with unrestricted fixed gears, which were generally chosen by the riders to be about 20 percent higher.

In extensive experimental work with ergometers at Loughborough University of Technology, one cyclist produced 1.5 hp (1,119 W) for 5 seconds. This is represented by the isolated point B on figure 2.7, and suggests that the gaps between the curves in that figure are likely to lessen as more ergometer experiments are done. Two curves in the NASA Bioastronautics Data Book (figure 2.13) give validity to these and other high-power data. The NASA book states that "data beyond one hour are sparse, and the maximum level one can sustain for 4–8 hours is not precisely known." [26] But these levels can be calculated from time-trial records, as in table 2.2. These long-duration data indicate that figure 2.13's curve for "first-class athletes" is somewhat low, but correct in its trend.

Figure 2.11

Power-output plots. Data for curves A, B, C, E, and F from reference 23; data for curve D from reference 21; curve G estimated by F.R.W. from data in reference 23; curve H extrapolated from data in reference 23. Peak efficiencies: 12.5% for curve A, 18% for curve B, 22% for curve C, 17% for curve D, 26% for curves E–G; optimum pedaling rates for the range of power outputs for curve H.

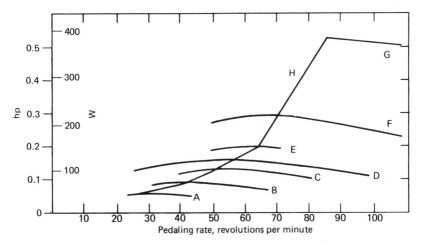

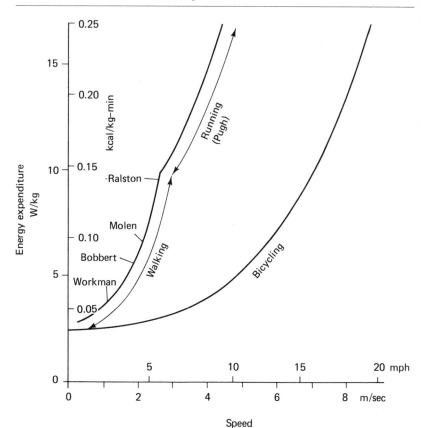

Speed

Figure 2.12
Energy expenditure in
kcal/kg-min during
walking, running, and
cycling, calculated from
formulas and from
experimental data of
Hermans-Telvy and
Binkhorst (reference 25).
Data on walking from N.
H. Molen and R. H.
Rozendal, Energy
expenditure in normal
test subjects walking on a
motor driven treadmill,
Kon. Nederl. Acad. Wet.
70 (1967): 192–200; H. J.

Ralston, Energy-speed
relation and optimal
speed during level
walking, *Int. Z. ang.
Physiol. einschl.
Arbeitsphysiol.* 17 (1958):
277–283; J. M. Workman
and B. W. Armstrong,
Oxygen cost of treadmill
walking, *J. Appl. Physiol.*
18 (1963): 798–803; A. C.
Bobbert, Energy
expenditure in level and
grade walking, *J. Appl.
Physiol.* 15 (1960): 1015–
1021. Data on running
from L. G. C. E. Pugh,

Oxygen intake in track
and treadmill running
with observations on the
effect of air resistance, *J.
Physiol.* 207 (1970): 823–
835; reference 25 (*).
Data on cycling from
reference 6 and from
reference 25.

Figure 2.13
Long-duration human
power output. Curves
from reference 26.

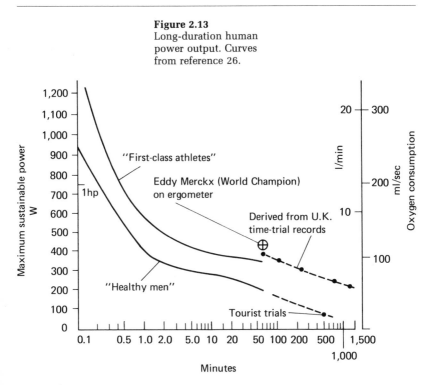

Table 2.2 British records for long-duration power output during bicycling.

Time trials, out-and-home, on the road

Event	Time	mph	m/sec	hp	W
25 miles (40.22 km)	49 min, 24 sec	30.4	13.6	0.5+	373+
50 miles (80.45 km)	103 min, 46 sec	28.9	12.9	0.5+	373+
100 miles (160.9 km)	225 min, 28 sec	26.6	11.9	0.44	325
1,000 miles (1,609 km)[a]	3,520 min	17	7.6	0.18	135

	Distance				
12 hours	281.9 miles (453.5 km)	23.5	10.5	0.37	276
24 hours	507 miles (815.8 km)	21.1	9.4	0.30	220

Tourist trials

100 miles (160.9 km)	8 hours	12	5.4	0.1	75

a. Records of the food consumption of J. Rossiter, who captured the 1,000-mile record in 1929, showed that he ate about 30 lb of eggs, milk, sugar, and chocolate during the 870 miles from Lands End to John O'Groats. No record of his drinks is available. The food energy content can be estimated as about 21,000 kcal during the approximately 2½ days of the ride. He had little rest, and his average riding speed was 6.7 m/sec (15 mph). Using this average speed, the distance traveled, reasonable figures for wind and friction losses, and the above metabolic efficiency, one would calculate an energy expenditure of 26,000 kcal. This is a fair agreement, considering the probability of body fat being consumed and the energy content of Rossiter's drinks. His energy output, in round figures, was thus 10,000 kcal/day, exceeding estimates of the maximum output of other athletes or of hard laborers by 100–200 percent.

Figure 2.14
Effects on maximum work of saddle height and angle of seat tube from perpendicular (a). 100 mm below normal, 21° from perpendicular; (b) 30 mm above, 8°; (c) 40 mm below, 21°; (d) 30 mm above, 43°; (e) normal height, 21°; (f) 30 mm above, 21° and 29°. From reference 27.

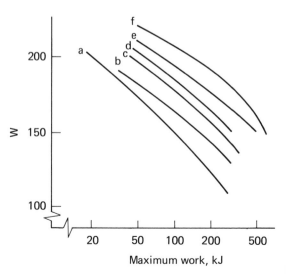

Effect of saddle height

Using a single subject (a 39-year-old man),
Müller[27] obtained the results shown in figure
2.14. He found that more power was obtained
when the saddle was raised by 40–50 mm (1.8–
2 in.) above the "normal" height (that for which
the heel can just reach the pedal with the leg
stretched and the posture upright). On the other
hand, minimum calorie consumption (or maximum energy efficiency) was found when the
saddle was lowered 40 mm below normal.

Thomas tested 100 subjects on a Müller ergometer, [28–30] and also found that maximum
power output was obtained with the saddle set
at a height about 10 percent greater than leg
length. He defined saddle height as the distance
from the pedal spindle at its lowest point to the
top of the saddle, so that about half of the thickness of the pedal would reduce the effective
height.

Effect of crank length

The safety bicycle has fixed the length of the
cranks within narrow limits. With the saddle at
the normal height above the pedals (as defined
by Müller), and with the pedals at a distance
above the ground such that in normal turns
(when the bicycle will be inclined toward the
center of the turn) the pedals will not contact
the ground, the saddle will be at a height at
which the rider can just put the ball of one foot
on the ground when stopped while still sitting
on the saddle. The crank length is then chosen
at a value at which almost all riders will feel
comfortable. This length is normally, for adult
riders, taken as 165 mm (6.5 in.) or 170 mm (6.7
in.). Thus, the height above the ground of the
bottom-bracket axle is fixed. An attempt to fit
longer cranks will lead to a reduction of pedal
clearance when cornering. Few riders, then,
have an opportunity to try long cranks, because
a specially designed frame is strictly necessary
for each crank length. (In this respect, recum-

bent bicycles have an advantage.) Most data on the effects of crank length have been taken on ergometers. Ergometer data can be regarded with suspicion, as we have implied, and this has certainly been true with regard to data on long cranks. So few people have been able to experiment with significantly longer cranks on actual bicycles (because special frames must be built) that their impressions also must be treated with reserve.

Two people writing for a bicycling magazine in 1897 advocated shorter cranks.[31] One, Perrache, experimented with 160-, 190-, and 220-mm cranks on a bicycle over a 5-km course and found that, in maximum-speed runs, he could get about 9 percent more power output with the 160-mm cranks than with the 220-mm cranks. We do not know whether the gear ratios were changed for different crank lengths. It would obviously penalize longer cranks if the gear ratio were not increased to give approximately similar ratios of pedal speed to wheel-rim speed. It would also be a disadvantage if the rider was accustomed to using short cranks.

Müller and Grosse-Lordemann tested the effect of crank lengths on an ergometer, employing only one subject.[32] Their approach was to set the power output the subject had to produce and to measure the maximum duration for which this output could be sustained. They also used three crank lengths: 140, 180, and 220 mm. In this case the subject was able to produce the most total work (that is, work for the longest periods) when using the longest cranks for all power levels. At the highest powers, the body efficiency (work output divided by energy input in food) was also highest when the longest cranks were used.

Harrison (ref. 15) gave his five subjects an initial choice of crank length, and found that they preferred the longer cranks (177 and 203 mm; 7 and 8 in.). The subjects were not particularly tall. Harrison intended to take all tests at two different crank lengths; however, he found from initial tests that "crank length played a relatively unimportant role in determining maximum power output," and used just one (unspecified) length for most of his tests.

The world champion Eddy Merckx used 175-mm cranks for the world's one-hour record, and has used 180-mm cranks for time trials and hill climbs in the Tour de France (ref. 31). A strong advocate of long cranks in the United States, James Farnsworth, uses them in achieving very fast climbs up Mount Washington.

In summary, crank length does not seem to be of major importance for producing moderate power outputs through pedaling.[33] The weight of the evidence on maximum power production is that longer-than-normal cranks (170–180 mm for normal-height adults), coupled with higher-ratio gearing to give similar foot speeds, give some advantages, at least in endurance. Figure 2.15 shows a pedal design that permits longer cranks.

Figure 2.15
Pedal design allowing greater crank length. Courtesy of Shimano American Corporation.

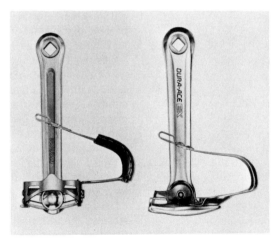

Effect of elliptical chainwheels

Elliptical chainwheels can be fitted to normal cranks in such a way that the pedal motion remains circular. The purpose is to reduce the supposedly useless time during which the pedals are near the top and bottom "dead centers." This topic has some similarity to that of long cranks in that there are fierce proponents and antagonists and few reliable data. Four of Harrison's five subjects produced virtually identical output curves (power versus time) using circular and elliptical chainwheels (ref. 15). One, apparently Harrison himself, gave about 12.5 percent more power with the elliptical chainwheel. All preferred the elliptical chainwheel for low-speed, high-torque pedaling. The degree of ovality was not specified, but Harrison stated that the foot accelerations required were high.

The degree of ovality can be specified by the ratio of the major to the minor diameter. (An illustration in Harrison's paper shows a chainwheel of about 1.45 ovality, which is a very high degree.) In the 1890s, racing riders using elliptic chainwheels with ovalities of about 1.3 became disillusioned with their performances, and these chainwheels fell out of favor. In the 1930s the Thetic chainwheel, with an ovality ratio of 1.1, became quite popular. Experiments with chainwheels having ovalities up to 1.6 confirm that high ovality (perhaps 1.2 or greater) decreases performance (F.R.W., unpublished). With a Thetic-type chainwheel, no deterioration of performance compared with that on a round chainwheel was recorded, and a small proportion of riders improved their performances by a few percent.

The parallel with the results for long cranks is striking. An elliptical chainwheel with a major-to-minor-diameter ratio of about 1.1, and cranks 5–10 percent longer than the present standard (coupled with a higher-ratio gear), sometimes give better performance, usually offer more comfortable pedaling, and apparently never diminish performance.

Effect of cam drive

Many people have invented and reinvented forms of linear drive in which the foot pushes on (for instance) a swinging lever, with a strap or cable attached to the lever at a point along it that can be varied to give different gearing ratios. The cable is then attached, perhaps through a length of chain, to a freewheel on the back wheel and to a return spring (figure 2.16). The drive of the American Star (figure 1.18) was of this type, although the gear was not variable.

The overwhelming disadvantage of this type of drive is that the feet and legs must be accelerated and subsequently decelerated by the muscles in the same way as in shadowboxing.[34] Harrison (ref. 15) found rather low outputs for motions of this type (figure 2.6, curves 2 and 4).

Pedalers of lever drives complain of the inability to use ankle motions for propulsion, as is possible with the common rotary drives. Some years ago in Germany a "foot cycle" was made for handicapped people. This machine, which could be propelled by the use of ankle motions only, demonstrated the help that the lower part of the legs can be to the ordinary bicyclist.

Figure 2.16
Swinging-lever (linear) drive.

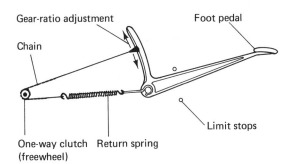

Gear-ratio adjustment

Foot pedal

Chain

Limit stops

One-way clutch   Return spring
(freewheel)

Two cam drives have been developed since 1975 by Lawrence Brown. The first, Beta II (which, over Brown's objections, became the Facet BioCam, shown in figure 2.17), combines the variable-gearing feature of linear drives with conservation of the kinetic energy of the moving masses and with conventional circular foot motion. Brown achieved this by fitting a cam in the place of the normal chainwheel and having the cam operate the lever system. By choosing the proper cam shape, he improved upon the elliptic chainwheel, making the foot and leg motion more suited to the optimum muscle action. Speed-distance records have been claimed for this drive system.

Figure 2.17
Bio-Cam drive. Courtesy
of Facet Cycle Inc.

Brown regarded Beta II as merely an interim stage, produced against the need to get a working machine to the 1978 Summer Olympics camp. He went on to design the Selectocam (figure 2.18), which brings the cranks quickly over the top and bottom dead centers and lengthens the duration of the high-thrust parts of the stroke. On the first race trial, in the 1980 Paris-Brest-Paris road race, Selectocam-equipped bicycles came in first, second, and third among American entries and second overall. Results of tests carried out by the American Sports Medical Training Center on four subjects riding conventional 10-speed lightweights and cycles equipped with the two Brown transmissions

Figure 2.18
Brown Selectocam drive. Courtesy of Lawrence G. Brown, Mechano-Physics Corp.

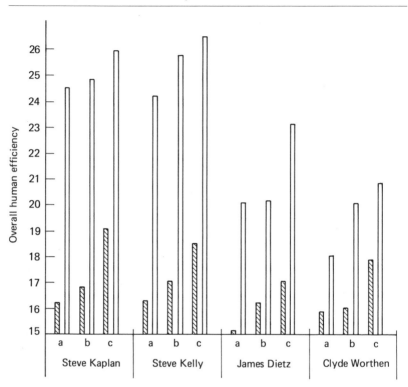

Figure 2.19
Comparison of overall efficiencies of four riders using conventional bicycles (shaded bars) and Beta II cam-drive bicycles (unshaded bars) with the following work loads and durations: (a) 134 watts, 0–5 minutes; (b) 134 watts, 5–10 minutes; (c) 163 watts, 10–15 minutes. From American Sports Medical Training Center.

mentioned here (there have been others) show impressive improvements in metabolic efficiency. Figure 2.19 is from a test on the Beta II.

These developments are significant. Two factors make these transmissions different: the "velocity profiles" of the foot motions and the very high gearing used. High gear in the successful French race mentioned above was 152 inches (3.85 m), 26 percent higher than normal high gears. Brown has now increased this to 185 inches (4.7 m), 53 percent higher than normal. We have drawn attention elsewhere in this book to other results which indicate that gears higher than normal would give higher efficiencies. However, enthusiastic riders of Selectocam bicycles have stated that the "natural" feeling of the foot rhythm makes them feel most comfortable in even higher gears than would seem optimum from tests.

Pedaling force

Table 2.3, compiled from data given in other parts of this book, compares the recorded pedaling rates of bicyclists of all types with estimates of the power outputs. These estimates, in turn, have led to estimates of the tangential forces at the pedals resisting the motion.

It appears that a paced bicyclist tends to use very consistent but moderate pedal thrusts, amounting to mean applied tangential forces of only about one-fifth of the rider's weight. The peak vertical thrusts are greater but still relatively small. No doubt this action enables the rider to maintain a steady seat position and to steer steadily.

It is easy to calculate from the crank length and the pedaling speed in revolutions per minute how much thrust upon the pedals is required for a given horsepower output. The peripheral pedal speed around the pedaling circle (or the vertical speed on the downward stroke) can be used in the equation

Thrust force (newtons)

$$= \frac{\text{Power (watts)}}{\text{Pedaling speed (m/sec)}} .$$

Ergometer experiments,[35] conducted under constant-speed pedaling conditions in order to check the agreement between the measured thrust and the calculated thrust, have shown that at the optimum pedaling speeds (related to power outputs as in figure 2.11) the measured thrust agreed with the predicted thrust to a reasonable accuracy, particularly for power outputs above 0.1 hp (74.6 W). At pedaling rates other than the optimum, the measured average vertical thrust upon the pedal over its path was greater than that expected by amounts that could be predicted from the lowering in pedaling efficiency as given on figure 2.11 by oxygen-

Table 2.3 Pedaling speeds.

	Distance (miles)	Time	mph	Gear (in.)	Crank (in.)	Crank speed (rpm)	Foot speed (ft/min)	Est'd power (hp)	Est'd thrust (lbf)
Ordinary, track	$\frac{1}{4}$	30 sec	30	53	5	190	493	1.35	91
	$\frac{1}{2}$	72 sec	25	56	5	150	392	1.05	88
		60 min	20.1	59	$5\frac{1}{2}$	116	330	0.5	50
Safety, track	$\frac{1}{8}$	12.4 sec	36.3	90	$6\frac{1}{4}$	136	446	1.6	120
	$\frac{1}{8}$	12.2 sec	37	68	$6\frac{1}{2}$	182	619	1.6	85
	$\frac{1}{4}$	29 sec	29.8	64	$6\frac{1}{4}$	170	520	1.3	83
	$\frac{1}{8}$	11.5 sec	39	90	$6\frac{1}{2}$	145	473	1.65	115
Safety, track, motorcycle paced		60 min	40.1	106	$6\frac{3}{4}$	126	445	0.5	37
		60 min	56	139	$6\frac{1}{2}$	134	456	0.5	36
		60 min	61.5	144	$6\frac{1}{2}$	143	488	0.5	35
		60 min	71	180	$6\frac{1}{2}$	133	454	0.5	36
		60 min	76	191	$6\frac{1}{2}$	134	454	0.5	36
Train-paced	1	57 sec	62	104	$6\frac{1}{2}$	198	670	1.2	59
Road safety bicycle	25	52 min	28.8	90	$6\frac{7}{8}$	102	370	0.6	54
	100	4 h	25	85	$6\frac{7}{8}$	99	368	0.5	45
	480	24 h	20	80	$6\frac{7}{8}$	84	310	0.25	26
	100	4 h 28 min	22.4	81	$6\frac{1}{2}$	93	316	0.5	52
Road, tourist			10	68	$6\frac{7}{8}$	50	180	0.09	16
			12	68	$6\frac{7}{8}$	61	220	0.11	16
			16	75	$6\frac{7}{8}$	74	266	0.2	24
			18.5	75	$6\frac{7}{8}$	85	305	0.3	32

Sources: A. C. Davison, Pedaling speeds, Cycling (20 January 1933): 55-56; H. H. England, I call on America's largest cycle maker, Cycling (25 April 1957): 326-327; 'Vandy,' the unbeaten king, Cycling (11 March 1964): 8; Marcel De Leener, Theo's hour record, Cycling (7 March 1970): 28.

consumption tests. Hence, it was concluded that, at other than optimum pedaling rates, thrust is "wasted" somewhere in the system. Maybe the body is lifted unnecessarily or the legs are swung so that lateral thrust components occur. At 60 rpm, measured pedal thrusts are near those expected from the ergometer power requirements. (Reference 36 also reported that professional bicyclists using pedals with toe straps did not use them to pull upward during the rising stroke (see also reference 37).

Bourlet[38] discussed a pedal made in 1897 by Bouny that made it possible to measure the vertical and horizontal components of pedal thrust (figure 2.20). One can see from figure 2.20 that much of the pedal thrust of this particular rider, particularly near bottom dead center, did not contribute to crank torque. The energy was wasted in merely lifting the rider's body. A. A. Zimmerman, the great American sprinter of the 1890s, was reported in *Cycling* (29 September 1894) as advocating that full leg thrust not be used at bottom dead center.

Figure 2.20
Magnitudes and directions of resultant forces at various points on the pedaling circle. Scale of foot forces is shown at lower left. Arrows show direction and magnitude of foot force on pedal; lines show angle of pedal platform.

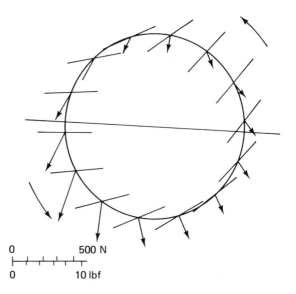

Bourlet was convinced of the value of dyna-mometer pedals for improving the efficiency of pedaling. However, he was severely critical of their use by French experimenters to compare the power needed to propel different combina-tions of machines and riders. He stressed the variability of the riders' frontal areas as they ad-justed their riding positions to suit various ped-aling postures. He preferred free-wheeling experiments, in which the riders' positions could be standardized.

A modern experimental version of the pedal-force diagram (figure 2.20) is given as figure 7-2 of reference 37.

Measurements made during actual bicycling

Thorough and accurate data relating oxygen consumption, heart rate, pedal torque, pedaling rate, bicycle speed, gear ratio, and crank length have been taken by the Japanese Bicycle Re-search Association by equipping several riders with instruments and telemetrically recording their behavior during actual riding (ref. 23). The results generally substantiate the foregoing dis-cussion. Figure 2.21 shows the relation between oxygen consumption and heart rate for four sub-jects ranging from a trained athlete to an every-day utility bicyclist. The best performance of one of the racing cyclists over various distances, using a range of gear ratios, is shown in figure 2.22. The best times and speeds were attained with the highest gear ratio—about 111 inches (2.82 m)—except for the shortest distance, 200 m, for which a range of gear ratios gave vir-tually identical average speeds. Tests of differ-ent crank lengths were inconclusive, but tended to show best performance with a crank length of $6\frac{3}{4}$ inches (approximately 170 mm) for the un-trained bicyclists and no significant effect of crank length on average speed over 1,000 me-ters for the trained riders.

An interesting cross-correlation of efficiency versus crank speed for various average speeds

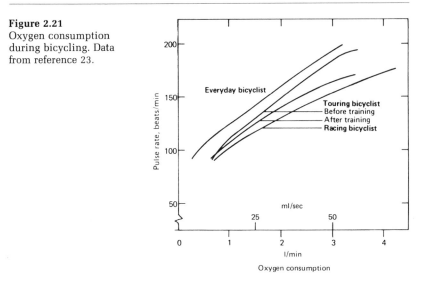

Figure 2.21
Oxygen consumption
during bicycling. Data
from reference 23.

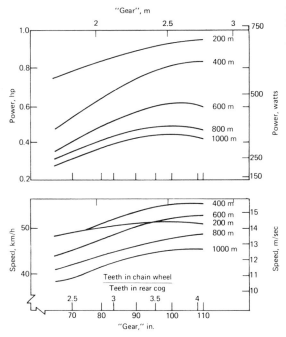

Figure 2.22
Effect of gear ratio on
performance of racing
bicyclist. From reference
23.

Figure 2.23
Effect of gearing on
energy efficiency. From
reference 23.

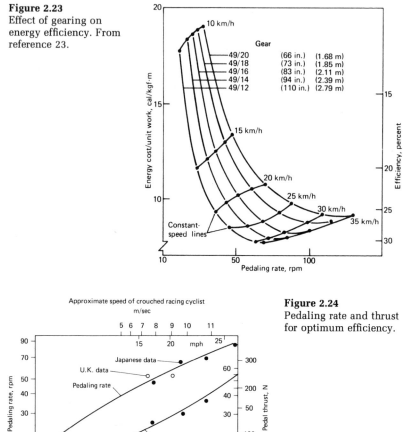

Approximate speed of crouched racing cyclist

Figure 2.24
Pedaling rate and thrust
for optimum efficiency.

and gear ratios is shown in figure 2.23 for the bicyclist who produced the most work per liter of oxygen consumption. The peak efficiency (about 30 percent) at the higher speeds (30 and 35 km/h; 8.33 and 9.72 m/sec) was obtained at 60–70 crank rpm and at the highest gear ratio of 111 inches (2.82 m).

A tentative conclusion is that racing bicyclists could use gear ratios higher than those usually employed, since peak efficiency was not reached even at 111 inches (2.82 m). Such gear sizes are coming into use as the top gears of multispeed bicycles, particularly for time-trial racing. The test riders did, however, complain a leg strain, and it may be advisable for most riders to continue with slightly lower gears giving pedaling rates of 90–100 rpm when pedaled to capacity.

References 6 and 14 summarize results obtained by earlier workers in actual measurements of oxygen breathing of bicyclists in motion. Figure 2.24, which cross-plots the data of reference 23 for maximum energy-use efficiency, shows both pedaling rate and pedal thrust increasing with power output. These data, taken during actual cycling, should be more reliable than those of reference 20, with which they disagree to some extent.

Conclusion

Pedaling as on conventional bicycles enables riders to approach their maximum power output. However, mechanisms that give noncircular foot motions or nonconstant velocities, or both, and mechanisms that allow hands and feet to be used together, seem to be required if the absolute maximum power output is to be obtained. Such mechanisms should make it possible to produce the lower outputs appropriate to utility bicycling in greater comfort than by conventional pedaling.

References

1. *Design and Use of Respirators* (Oxford: Pergamon, 1961).

2. W. von Döbeln, A simple bicycle ergometer, *Journal of Applied Physiology* 7 (1954): 222–224.

3. C. Lanooy and F. H. Bonjer, A hyperbolic ergometer for cycling and cranking, *Journal of Applied Physiology* 9 (1956): 499–500.

4. C. T. M. Davies and J. M. Musgrove, The aerobic and anaerobic components of work during sustained exercise on a bicycle ergometer, *Ergonomics* 14 (1971), no. 2: 237–263.

5. C. H. Wyndham et al., Inter- and intra-individual differences in energy expenditure and mechanical efficiency, *Ergonomics* 9 (1966), no. 1: 17–29.

6. F. R. Whitt, A note on the estimation of the energy expenditure of sporting cyclists, *Ergonomics* 14 (1971), no. 3: 419–424.

7. H. B. Falls, *Exercise Physiology* (New York: Academic, 1968).

8. L. Brouha, *Physiology in Industry*, second edition (Oxford: Pergamon, 1967).

9. G. H. G. Dyson, *The Mechanics of Athletics* (University of London Press, 1962).

10. T. Vaughn, *Science and Sport* (London: Faber and Faber, 1970).

11. A. W. Hill, *Trails and Trials in Physiology* (London: Clowes, 1965).

12. E. A. Müller, Physiological methods of increasing human work capacity, *Ergonomics* 8 (1965), no. 4: 409–424.

13. H. W. Knipping and A. Moncrieff, The ventilation equivalent of oxygen, *Queensland Journal of Medicine* 25 (1932): 17–30.

14. W. C. Adams, Influence of age, sex and body weight on the energy expenditure of bicycle riding, *Journal of Applied Physiology* 22 (1967): 539–545.

15. J. Y. Harrison, Maximizing human power output by suitable selection of motion cycle and load, *Human Factors* 12 (1970), no. 3: 315–329.

16. R. C. Carpenter, et al., The relationship between ventilating capacity and simple pneumonosis in coal workers, *The British Journal of Industrial Medicine* 13 (1965): 166–176.

17. L. G. C. E. Pugh, The relation of oxygen intake and speed in competition cycling and comparative ob-

servations on the bicycle ergometer, *Journal of Physiology* (1974): 795–808.

18. T. Nonweiler, The work production of man: Studies on racing cyclists, *Proceedings of the Physiological Society* (11 January 1958): 8P–9P.

19. C. R. Kyle, V. J. Caizzo, and M. Palombo, Predicting human powered vehicle performance using ergometry and aerodynamic drag measurements, Conference on Human Power for Health, Productivity, Recreation and Transportation, Technology University of Cologne, September 1978.

20. H. Grosse-Lordemann and E. A. Müller, Der Einfluss der Leistung und der Arbeitsgeschwindigkeit auf das Arbeitsmaximum und den Wirkungsgrad beim Radfahren, Kaiser Wilhelm Institut für Arbeitsphysiologie, Dortmund-Munster, 1936.

21. R. C. Garry and G. M. Wishart, On the existence of a most efficient speed in bicycle pedalling and the problem of determining human muscular efficiency, *Journal of Physiology* 72 (1931): 425–437.

22. Report on the Energy-Storage Bicycle, Thayer School of Engineering, Dartmouth College, Hanover, N.H., 1962.

23. Report of the Bicycle Production and Technical Institute, Japan, 1968.

24. D. R. Wilkie, Man as an aero-engine, *Journal of the Royal Aeronautical Society* 64 (1960): 477–481.

25. E. J. Hermans-Telvy and R. A. Binkhorst, Lopen of Fietsen?—Kiezen op Basis van Het Energieverbruik, *Hart Bulletin* (Netherlands), June 1974.

26. U.S. National Aeronautics and Space Administration, Bioastronautics Data Book, document SP-3006 (1964).

27. E. A. Müller, Der Einfluss der Sattelstellung auf das Arbeitsmaximum und den Wirkungsgrad beim Radfahren, Kaiser Wilhelm Institut für Arbeitsphysiologie, Dortmund-Munster, 1937.

28. V. Thomas, Saddle height, *Cycling* (7 January 1967): 24.

29. V. Thomas, Saddle height—Conflicting views, *Cycling* (4 February 1967): 17.

30. E. J. Hamley and V. Thomas, The physiological and postural factors in the calibration of the bicycle ergometer, *Journal of Physiology* (1967): 191.

31. F. DeLong, *DeLong's Guide to Bicycles and Bicycling* (Radnor, Pa.: Chilton, 1978).

32. E. A. Müller and H. Grosse-Lordemann, Der Einfluss der Tretkurbellange auf das Arbeitsmaximum und den Wirkungsgrad beim Radfahren, Kaiser Wilhelm Institut für Arbeitsphysiologie, Dortmund-Munster, 1937.

33. F. R. Whitt, Crank length and pedalling efficiency, *Cycling Touring* (December–January 1969): 12.

34. S. S. Wilson, Bicycling technology, *Scientific American* (November 1973): 81–91.

35. F. R. Whitt, Pedalling rates and gear sizes, *Bicycling* (March 1973): 24–25.

36. M. J. A. Hoes et al., Measurement of forces exerted on pedal and crank during work on a bicycle ergometer at different loads, *Internationale Zeitschrift für angewandte Physiologie einschliesslich Arbeitsphysiologie* 26 (1956): 33–42.

37. I. E. Faria and P. R. Cavenagh, *The Physiology and Biomechanics of Cycling* (New York: Wiley, 1978), p. 91, figure 7-2.

38. C. Bourlet, II nouveau traité des bicycles et bicyclettes, le travail (Paris: Gauthier-Villars, 1896).

Recommended reading

J. D. Brooke and G. J. Davies, Problems in the use of respiratory variables to assess field work demands and to replicate them in laboratory tasks, in *Lung Function and Work Capacity* (Salford University, U.K., 1970).

J. D. Brooke and M. S. Firth, A machine for testing and training cyclists, *British Cycling Federation Coaching News* (March 1972).

———, Calibration of a simple eddy current ergometer, *British Journal of Sports Medicine* 8 (1974), no. 2/3: 120–125.

J. M. Hagberg et al., Comparison of three procedures for measuring VO²max in competitive cyclists, *European Journal of Applied Physiology* 39 (1978): 47–52.

S. B. Stromme et al., Assessment of maximal aerobic power in specifically trained athletes, *Journal of Applied Physiology* 42 (1977): 833–837.

C. Juden, Oval-Chainwheel Design, Part II, project thesis, Churchill College, Cambridge University, April 1977.

T. A. McMahon and P. R. Greene, Fast running tracks, *Scientific American* (December 1978): 112–121.

A. B. Streng, Ondersoek ann de Aandrijving van door Spierkracht voortbewogen Twee-Wielige Voertuigen, thesis, Technische Hogeschool Twente.

3   How bicyclists keep cool

Bicycling can be hard work. It is important that the body, like any engine, not become overheated when producing power. We pointed out in chapter 2 that the measurement of the power output of bicyclists on ergometers is open to criticism because the conditions for heat dissipation are critically different from those occurring on bicycles. The performances of riding bicyclists in time trials are, however, very amenable to analysis. Such time trials are of far longer duration than the few hours usually assumed (see, for example, reference 1) as the maximum period over which data on human power output are available. Time trials (unpaced) are regularly held for 24-hour periods; distances of 480 miles (772 km) are typical.

The air blast generated by bicycling is of such magnitude that it bears little resemblance to the drafts produced by the small electric fans sometimes advised for cooling pedalers on ergometers. As a consequence it can be said that under most conditions of level cycling the bicyclist works under cooler conditions than does an ergometer pedaler. At high speeds, most of the rider's power is expended in overcoming air resistance. At 20 mph (8.94 m/sec) about 0.2 hp (149 W) is dissipated in the air. The cooling is a direct function of this lost power. Even if the little fans often used for ergometer experiments ran at this power level, the cooling effect would be much less than that for the moving bicyclist, because little of this power is dissipated as air friction around the subject's body.

The science of "convective" heat transfer between a surface and a gas in relative motion is based largely on the analogy between fluid friction and conduction heat transfer derived by Osborne Reynolds in 1874 (ref. 2, pp. 134–137).

Reynolds's analogy states that the heat transferred between the body surface and the air flowing past is proportional to the air friction at the surface multiplied by the difference of temperature between the surface and the air. Therefore, we can think of surface air friction as partly useful, at least in warm weather. Much of the air friction that slows a bicyclist occurs as eddies in the wake behind the body. These do not contribute to heat transfer in any way.

The effect of adequate cooling may be inferred from Wilkie's finding from experiments with ergometer pedalers that if any capability of exceeding about half an hour's pedaling is required, the subject must keep his power output down to about 0.2 hp (149 W).[3] However, peak performances in 24-hour time trials can be analyzed using wind-resistance and rolling-resistance data from reference 1 to show that about 0.3 hp (224 W) is being expended over that period (see table 2.2). It seems that the exposure of the pedaler to moving air is principally responsible for the improvement. It is also known that an ergometer pedaler who attempts a power output of about 0.5 hp (373 W) can expect to give up after some 10 minutes and will be perspiring profusely. That is the same power output required to propel a racing bicyclist doing a "fast" 25-mile (40.2-km) distance trial of nearly one hour. Again the effect of moving air upon a pedaler's performance is very apparent.

Let us examine the literature for suitable correlations of established heat-transfer data in order to find quantitative reasons for the above observations.

Heat-transfer data and deductions

Because there is no published information concerning heat-transfer experiments with ridng bicyclists, it is necessary to make calculations with suitable approximations of a cyclist's shape. The approximate forms used are a flat plate and a 6-inch-diameter cylinder. In addition, data from experiments with actual human forms (refs. 3–5) can be looked at, although the flat and upright postures used were not those of bicycling.

The results of many calculations using established correlations for both convective and evaporative heat transfer are given in figure 3.1. Also shown is the heat evolution of a rider at various speeds and power outputs on the level. The figure indicates that the effect of shape on the flux for a given temperature difference is not great in the case of convective heat transfer. In the case of evaporative transfer, the difference between results with models and with an actual human body is 20 percent. It appears that a midway value can be obtained from data on crossflow over wetted 6-inch-diameter cylinders or plates. For the same driving potential, expressed as water-vapor pressure or temperature difference, the rate of evaporative heat transfer is about double that of convective transfer.

Under normal free-convection conditions, data given in references 6 and 7 lead to the conclusion that cooling is performed by air moving at about 1.5 ft/sec (0.457 m/sec). This is supported by figure 3.1, where line 6 for forced convection over a cylinder at 1.5 ft/sec and point 9 for free convection both predict about 325 watts per square meter as the heat flux for that air speed. This value would be greater for a bicyclist, whose legs would also be moving the air.

In the design of heating and ventilating plants, the maximum heat load produced by a worker doing hard physical labor has long been accepted as 2,000 Btu/h (586 W).[8,9] This figure, when applied to a body surface of 1.8 m², also amounts to 325 W/m². It is recommended that such hard work be done at a room temperature of 55°F (12.8°C). Most of the heat is lost through evaporation of sweat.

The above evidence leads to the conclusion that a rider pedaling in such a manner that his body gives out a total of 2,000 Btu/h (586 W), in

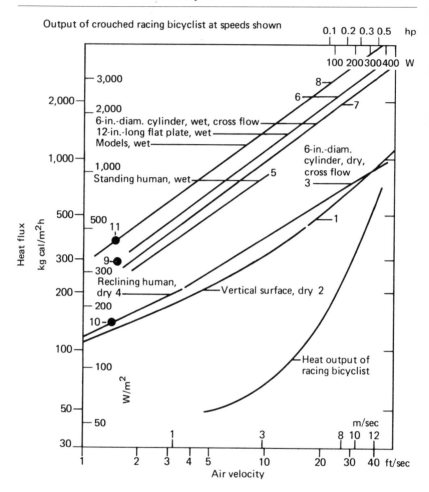

Figure 3.1
Convective and evaporative heat flows. Assumed conditions: surface temperature 35°C (constant), air temperature 15°C (constant), relative humidity of air 80 percent. Data for curves 1 and 2 from reference 6, p. 857; data for curve 3 from W. H. McAdams, *Heat Transmission* (New York: McGraw-Hill, 1942), p. 223; data for curves 4 and 8 from reference 4, p. 37; data for curve 5 from reference 5, p. 257; data for curves 6 and 7 and points 9 and 11 from C. Strock, *Heating and Ventilating Engineer's Databook* (New York: Industrial, 1948), pp. 5–12. Data for curve giving heat output of racing bicyclist are from metabolic-heat data adjusted for mechanical power and some small heat-energy equivalent. Bicyclist's body surface is assumed to be 1.8 m². See table 2.1.

average air conditions where free convection holds, does not suffer a noticeable rise in body temperature no matter how long he works. If the pedaler's efficiency is 25 percent, the work output W for a heat loss Q is calculated as follows:

$$0.25 = \frac{W}{W + Q} = \frac{1}{1 + Q/W} \; ;$$

therefore

$$W = Q/3 = 195 \text{ watts} = 0.26 \text{ hp.}$$

Thus, it seems that a pedaler on an ergometer working for long periods produces only about 0.2 hp (149 W) because of unwillingness to tolerate a noticeable rise in body temperature.

In chapter 2 it was shown that many cyclists can exert 0.5 hp (373 W) for periods of up to an hour. According to figure 2.24, that corresponds to a speed of about 27 mph (12.2 m/sec). At that speed the heat flow from the moving bicyclist is about 707 W/m^2 (fig. 3.1). If the cyclist exerts 0.5 hp (373 W) pedaling on an ergometer, all the heat lost by convection and evaporation in moving air—all of the heat in excess of 325 W/m^2—must be absorbed by the pedaler's body. Thus, the ergometer pedaler with a body area of 1.8 m^2 absorbs $(707 - 325) \times 1.8 = 688$ W if the small heat losses through breathing are neglected.

If the pedaler weighs 70 kilograms and has a specific heat of 3.52 joules per gram per °C, and if a rise in body temperature of 2°C is acceptable before physical collapse, the tolerable time limit for pedaling is

$$\frac{70{,}000 \times 3.52 \times 2}{688 \times 60} = 12 \text{ minutes.}$$

For highly trained racing bicyclists attempting to pedal ergometers at a power output of 0.5 hp (373 W), a common range of endurance is 5–15 minutes (personal observations, F.R.W.). Hence, the above estimates have some validity. The fact that all the racers observed were capable of outputs of nearly 0.5 hp (373 W) in one-hour time trials demonstrates vividly the value of flowing air in prolonging the tolerable period of hard work.

Experimental findings supporting the vital importance of cooling in human-power experiments are given in a paper concerning the effect of heat upon the performances of ergometer pedalers.[10]

Minimum air speed

From figure 3.1 it can be seen that a racing bicyclist producing 0.6 hp (450 W) evolves heat at about 850 W/m^2. According to curve 5, such a heat rate could be absorbed by air moving at about 3 m/sec (7 mph). Verification of the value of this prediction is found in Bill Bradley's performance on the Gross Glockner hill climb. He rode at about 5.4 m/sec (12 mph) at a power output of 450 W (0.6 hp), demonstrating that it is not necessary to have a road speed above 12 m/sec (27 mph) for level riding at 450 W, when nonevaporative heat transfer can cool if the air is at a lower temperature than the body. Bradley's ride was done in high-air-temperature conditions, but these were well compensated for by the low humidity of about 40 percent; he could sweat freely and achieve efficient evaporative heat loss.

Chester Kyle and co-workers at California State University carried out extensive trials with streamlined casings for riders of various machines.[11] An interesting outcome was that, even in short rides, a casing that was skirted almost to ground level caused the rider to overheat grossly—almost certainly because of a lack of air flow through the casing. This problem seems to have been appreciated even with the earliest bicyclist casings, developed in the early 1900s. Bryan Allen also suffered from overheating in the pedaled airplane *Gossamer Albatross* because of insufficient through-ventilation and insufficient water during the nearly three-hour flight across the English Channel.

Bicycling in cold and hot conditions

A problem faced by advocates of bicycling as a means for daily commuting is that even temperate regions have days, and sometimes weeks, of extreme weather conditions during which bicycling may be unpleasant for many and impossible for some.

There is no one set of temperature boundaries at which bicycling becomes impossible. Many "fair-weather" cyclists put their machines away for the winter when the morning temperatures drop to 10°C (50°F), and will not ride in business clothes at temperatures above 25°C (77°F). However, many hardier folk find bicycling to be still enjoyable at −15°C (5°F). The main problem at temperatures below this seems to be the feet. The size of insulated footwear is limited to that which can fit on bicycle pedals, and it is fairly common experience that, at −18°C (0°F), even when the trunk of the body is becoming overheated through exertion the feet can become numb with cold.

The effects of cold air are intensified by wind. Weather forecasters often express these effects in terms of "wind-chill factors"—the air temperatures that would have to exist, without wind, to give the same cooling to a human body as the combination of actual temperature and actual relative wind. The wind-chill factors tabulated by the U.S. National Oceanic and Atmospheric Administration are plotted in figure 3.2. With this chart, one can find the effect of bicycling into a relative wind. For instance, if the air temperature is −18°C (0°F) and one is bicycling into a relative wind of 5 m/sec (11 mph), the cooling is the same as if one were in calm conditions at a temperature of −30°C (−22°F). The feet are periodically traveling at a higher relative velocity (as they come over top dead center) and then at a lower velocity relative to the wind. Because the cooling relationship is nonlinear, the average effect seems to be more severe.

At the higher temperatures, humidity becomes very important. The bicycle is highly prized for

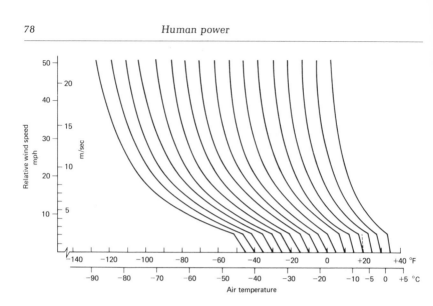

Figure 3.2
Wind-chill factors.
Plotted from National
Oceanic and
Atmospheric
Administration data.

personal transportation and for local commerce
throughout Africa and Asia. In northern Nigeria,
for example, the air is so dry throughout most
of the year that one's range on a bicycle is limited more by the availability of water than by
the temperature. The long-distance bicyclist Ian
Hibell rode through the Sahara (principally at
night), limited again by his water supplies. He
could not carry sufficient water for the longer
stages between oases, and relied on gifts of
water from passing motor travelers.

During the record heat wave of July 1980,
Houston, Texas had over four weeks of temperatures over 100°F (38°C), coupled with very high
humidity, with 111°F (44°C) reached on several
occasions. Yet some bicyclists continued to ride
to work. What makes this especially remarkable
is that on American roads, crowded with cars,
trucks, and buses with air conditioners going at
their maximum, the ambient temperature which
bicyclists must experience can be far above the
local off-highway values.

There are three lessons to be learned from the
experience of the hardier riders who brace what

seem to be extreme conditions. First, the promotion of good circulation through exertion helps the body cope with high temperature and high humidity as well as with cold weather. Second, the relative airflow which bicycling produces is a major factor in making riding in hot weather tolerable and usually enjoyable. Third, the fact that so many riders choose to bicycle in extreme conditions (rather than being forced to do so by economic necessity) shows that many other healthy but more timid cyclists could push out their limits without fear of harm.

Physiology of body-temperature regulation

Reference 12 includes a survey of recent experimental work on the complex processes involved in body temperature regulation, and a large bibliography.

Heat-transfer comparison of swimming, running, and bicycling

Swimmers are believed to maintain 65 percent of top velocity for one hour; runners only 55 percent. This is deducible from figure 7.1, which also shows that bicyclists maintain even higher degrees of efficiency than swimmers.

Water is a far better heat-removal fluid than air. Thus, with appropriate water temperatures, a swimmer can keep cool more easily than a runner. These conclusions are summarized in the statement that the swimmer uses a smaller proportion of the cardiac output to dissipate heat, and a larger proportion to transport oxygen to the muscles, than a runner. This statement appears to be just as appropriate to a bicyclist as to a swimmer, in comparison with a runner.

Conclusions

The heat-removal capacity of the air surrounding a working human is a key factor in the duration of his effort. Static air conditions are apparently such that, at low air speeds with free convection, the air is capable of removing 2,000 Btu/h (586 W) from the average body surface. Hence, if more heat is given out from working at rates higher than about 0.2 hp (149 W), the body temperature rises. (An ambient temperature of 55°F, or 12.8°C, is assumed.)

The fast-moving air around a bicyclist traveling on the level can be estimated to have a heat-removal capacity much greater than that of the stationary air surrounding an ergometer pedaler. Quantitative estimates can be made using established heat-transfer correlations based on air flow over wet 6-inch-diameter cylinders (cross-flow)[13] or from data given in reference 5 on air flow over a standing perspiring person.

The heat-removal capacity of the air around a moving cyclist, at most speeds on the level, is such that much more heat can be lost than the amount produced by the cyclist's effort. Hence, a rider can wear more clothing than the amount that would be tolerable to a static worker giving out the same mechanical power.

Some speculations

At least two ergometers used for testing the power capacities of racing bicyclists have incorporated air brakes in the form of fans. However, no one to date appears to have thought of directing the air from such air brakes onto the body of the pedaler and measuring the effect of the fast-moving air on performance. It is improbable that an air flow from such an arrangement could give anything very far from, say, half the flow rates surrounding an actual riding bicyclist giving out the same power, but the results would be interesting.

Pedaling on an ergometer out of doors should result in higher power output. Even in calm conditions, air is likely to be moving faster than the $1\frac{1}{2}$ ft/sec (0.457 m/sec) quoted above for free-convection conditions around a heated body.

In view of the fact that, at 0.2 hp (149 W) output, for tolerable body temperatures the body must get rid of its heat by an evaporative process, indoor exercise seems rather unhealthy compared with riding a bicycle in the open air.

Maybe the designers of home exercisers should put less emphasis on instrumentation and more on self-propelled cooling equipment.

References

1. D. R. Wilkie, Man as an aero-engine, *Journal of the Royal Aeronautical Society* 64 (1960): 477–481.

2. E. R. G. Eckert, *Introduction to Heat and Mass Transfer* (New York: McGraw-Hill, 1963).

3. T. Nonweiler, Air Resistance of Racing Cyclists, report 106, College of Aeronautics, Cranfield, England, 1956.

4. J. Colin and Y. Houdas, Experimental determination of coefficient of heat exchanges by convection of the human body. *Journal of Applied Physiology* 22 (1967), no. 1: 31–38.

5. D. Clifford, D. McKerslake, and J. L. Weddell, The effect of wind speed on the maximum evaporative capacity in man, *Journal of Physiology* 147 (1959): 253–259.

6. J. R. Perry, *Chemical Engineers Handbook* (New York: McGraw-Hill, 1936), pp. 339, 958–965.

7. R. N. Cox and R. P. Clarke, The natural convection flow around the human body, *Quest* (City of London University), 1969, pp. 9–13.

8. *Kempe's Engineer's Year Book,* vol. 11 (London: Morgan, 1962), pp. 761, 780.

9. O. Faber and J. R. Kell, *Heating and Air Conditioning of Buildings* (Cheam, England: Architectural Press, 1943).

10. C. G. Williams, et al., Circulatory and metabolic reactions to work in heat, *Journal of Applied Physiology* 17 (1962): 625–638.

11. C. R. Kyle, The aerodynamics of man-powered land vehicles, Third National Seminar on Planning, Design, and Implementation of Bicycle and Pedestrian Facilities, San Diego, Calif., 1974.

12. H. B. Falls, *Exercise Physiology* (New York: Academic, 1968).

13. T. K. Sherwood and R. L. Pigford, *Absorption and Extraction* (New York: McGraw-Hill, 1952), pp. 70, 87–89.

Recommended reading

A. Hardy, Warmth, *Bicycling* (December 1975): 22–23.

O. G. Edholm, *The Biology of Work* (Weidenfeld and Nicolson, 1967).

A. B. Craig, *Journal of Sports Medicine and Physical Fitness* 3 (1963): 14.

Daily Mail (England), World champion Hugh Porter drops only 3.2 seconds over 5 kilometres wearing Trevira jersey, and clothes, 9 May 1973. See also 8, 11, and 12 May issues.

W. H. Rees, Clothing and comfort, *Shirley Link* (Shirley Institute, Manchester) (summer 1969): 6–9.

E. R. Nadel, *Problems with Temperature Regulation During Exercise* (New York: Academic, 1977).

SOME BICYCLE PHYSICS

4 Wind resistance

"Wind resistance" is an everyday experience, particularly to bicyclists. It is caused by two main types of forces: one normal to the surface of the resisted body (felt as the pressure of the wind) and the other tangential to the surface (which is the true "skin friction"). For an unstreamlined body, such as a bicycle and rider, the pressure effect is much the larger, and the unrecovered pressure energy appears in the form of eddying air motion at the rear of the body. Part a of figure 4.1 shows this eddying effect at the rear of a cylinder. As can be seen in part b, the streamlined shape produces less eddying than the cylinder.

Vehicles intended for high speeds in air are almost always constructed to minimize eddying. Streamlined shapes incorporate gradual tapering from a rounded leading edge. The exact geometry of shapes that maximize the possibility of the flow remaining attached (rather than eddying) and minimize the skin friction can be approximated by rather complex mathematics. It is usual in aeronautics either to refer to one of a family of published "low-drag" shapes or to test models in a wind tunnel.[1,2]

The measurement of wind resistance of motor vehicles is described in reference 3. Although wind-tunnel experiments can yield good data for motor vehicles, the interaction of the air flow around the bicyclist with the "moving" ground is relatively more important for bicyclists. This reduces the validity of wind-tunnel data on bicyclists. More accurate information can be obtained with actual riders.

One aim of aerodynamic experiments on an object is to measure its drag coefficient C_D, defined as the nondimensional quantity

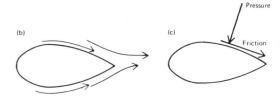
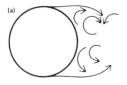

Figure 4.1
Effects of bluff and
streamlined shapes:
(a) eddying flow around
circular cylinder;
(b) noneddying flow
around streamlined
shape;
(c) pressure recovery
possible in the absence of
eddies.

$$C_D \equiv \frac{\text{Drag force}}{\text{Dynamic pressure of air} \times \text{Frontal area}}. \tag{4.1}$$

At low speeds (say, below 100 mph or 45 m/sec), the dynamic pressure is given by

$$\frac{\text{Air density} \times (\text{Relative velocity})^2}{2g_c},$$

where g_c is the constant ($=32.174$ lbm ft/lbf sec^2) in the equation $F = ma/g_c$, which relates pounds mass (m), pounds force (F), and acceleration (a, in ft/sec^2) through Newton's first law; and where the relative velocity is the velocity of the air moving past the object. In S.I. units, $g_c = 1.0$; m is in kg, F in newtons, and a in m/sec^2. Thus, the drag force is

$C_D \times$ Air density $\times$ (Relative velocity)^2
$\times$ Frontal area/$2g_c$.

The propulsion power necessary to overcome drag is

$P =$ Drag force $\times$ Relative vehicle velocity.

Since the drag force is approximately proportional to the square of the velocity, the power to overcome drag is approximately proportional to the cube of the velocity.

Only in still air is the vehicle velocity the same as the relative velocity used to calculate the drag force. When there is a headwind or a tailwind, the relative velocity is different from the vehicle velocity.

If the drag is measured in pounds force and the velocity is given in feet per second, the power is in ft-lbf/sec. This may be converted to horsepower by dividing by 550 (1 hp = 550 ft-lbf/sec); or miles per hour (1 hp = 375 mile-lbf/h) may be used:

$$P \text{ (hp)} \equiv \frac{\text{Drag (lbf)} \times \text{Velocity (ft/sec)}}{550 \text{ (ft-lbf/sec)/hp}}$$

$$\equiv \frac{\text{Drag (lbf)} \times \text{Velocity (mph)}}{375 \text{ (mile-lbf/h)/hp}}.$$

In S.I. units the relationship is

$$P \text{ (watts)} \equiv \text{Drag (newtons)} \times \text{Velocity (m/sec)}.$$

Drag

The drag coefficients of bodies whose drags are almost entirely due to pressure drag are virtually constant, whatever the conditions. (Examples of such bodies are thin plates set normal to the direction of flow.) But bodies with substantial contributions from the surface-friction drag of the so-called boundary layer of "sticky" or viscous flow have drag coefficients that can vary widely in different circumstances. In general, the flow in this boundary layer can exist in one of three forms: laminar, in which the layers of fluid slide smoothly over one another; turbulent, in which the boundary layer is largely composed of small confined but intense eddies which greatly increase the surface friction; and separated, in which the boundaries layer leaves the surface and usually breaks up into large-scale unconfined eddies.

If we wanted to produce a low-drag bicycle enclosure, we would prefer that the boundary-layer flow be entirely laminar (airplane designers have tried to arrive at laminar-flow wings). Unfortunately, laminar-flow boundary layers are extremely sensitive. They have a strong tendency to separate from the surface, producing very high drag. Turbulent boundary layers have higher surface friction than laminar boundary layers, and therefore give somewhat higher drag; however, they are less likely to separate. Often the lowest integrated drag is produced by forcing the laminar boundary layer on the forward part of a body to become turbulent, which at low speeds may require either the roughening of the surface or the mounting of a "trip" wire, at well before the location where separation might otherwise occur. A classic experiment by the aerodynamics genius Ludwig Prandtl showed this effect graphically. Prandtl mounted a smooth sphere in an airstream, measured its drag, and observed the airflow with streams of smoke. The flow separated even before the maximum diameter was reached (figure 4.2, top), and the drag was high. Then he fastened a thin wire ring as a boundary-layer trip on the upstream part of the sphere. The flow remained attached over a much larger proportion of the sphere's surface (figure 4.2, bottom), and the drag decreased greatly, as can be seen from the much smaller wake. Manufacturers of golf balls learned from this and roughened the surface with sharp-edged dimples, producing balls that could be driven faster and farther. (The dimples, combined with "top spin," also produce an aerodynamic lift force, which contributes to increasing the range.)

For any one shape of body, the variable that controls the drag coefficient is the Reynolds number, defined for a sphere moving in air as

$$\frac{\text{Air density} \times \text{Sphere diameter} \times \text{Relative velocity}}{\text{Air viscosity}} \quad (4.3)$$

For air at sea-level pressure and 65°F (19°C), this becomes

$$\tfrac{2}{3} \times \text{Sphere diameter} \times \text{Relative velocity} \times 10^5, \quad (4.4)$$

where the diameter is in meters and the velocity is in m/sec. At Reynolds numbers over 3×10^5, even smooth spheres do not need trip wires or rough surfaces (as can be seen from figure 4.3),

Figure 4.2
Effect of roughness on
drag of a smooth sphere
(Prandtl's experiment).
From S. Goldstein,
*Modern Developments in
Fluid Dynamics* (London:
Oxford University Press,
1938).

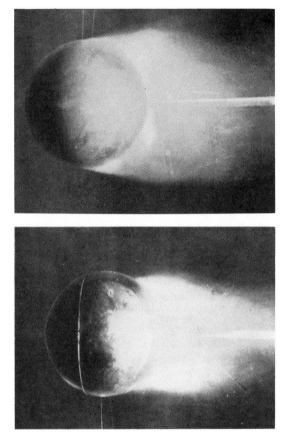

because a laminar boundary layer will spontaneously become turbulent under these conditions. When the boundary layer becomes turbulent, the drag coefficient falls sharply from 0.47 to 0.10. A golf ball about 40 mm in diameter driven at an initial velocity of 75 m/sec has a Reynolds number of 2×10^5 at the start, and would be in the high-drag-coefficient region if it were smooth. The dimpling shifts the "transition" point to lower Reynolds numbers and gives a low C_D. Thus, paradoxically, a rough surface can lead to low drag.

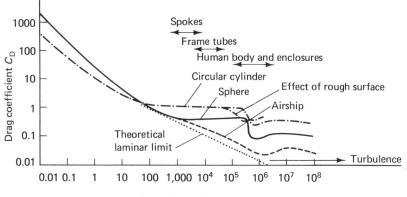

Figure 4.3
Drag coefficients of simple shapes. Data from reference 2. (The airship considered here is an R100A type with a length/diameter ratio of 5, which is typical of streamlined bodies.)

Compared with a golf ball, a bicyclist travels much slower but has a larger equivalent diameter, so the Reynolds number may be similar. A bicyclist using an upright posture may be considered for simplicity as a circular cylinder normal to the flow, a curve for which is shown in figure 4.3. If the equivalent diameter is 600 mm and the speed is 5 m/sec, the Reynolds number is 2×10^5—below the "transition" region of about 4×10^5. There may be some advantage to wearing rough clothing for speeds in this region. Most bicyclists have become aware of the penalty of converting themselves into smooth bodies by donning a wet-weather cape or poncho, which usually greatly increases the wind resistance without increasing the cross-sectional area. Perhaps some "trips" woven into the cape material would be beneficial. Even better would be some type of frame which would convert the cape into a low-drag shape. Sharp proposed this step in 1899,[4] and capes with inflatable rims were for sale around that time.

Low-drag shapes do not generally exhibit the sharp transition from high drag (separated flow) to low drag (attached flow) as the Reynolds number is increased. Rather, the point of transition of the boundary layer from laminar to tur-

bulent tends to move upstream toward the
leading edge of the body as the Reynolds number is increased. Thus, the drag coefficients for
streamline shapes given in figure 4.3 show a
continuous fall as the Reynolds number is increased in the laminar-flow region, followed by
a moderate rise to the fully turbulent conditions
and then a continued fall.

The Reynolds numbers of streamlined shells
for human-powered vehicles lie in the interesting transition region between 1.5×10^5 and
1.5×10^6.

To reduce the wind-induced drag of a bicycle
and rider, two alternatives are to reduce the
frontal area of rider plus machine and to reduce
the drag coefficient which the combined body
presents to the airstream. For years, bicyclists
have adopted one or other of these possibilities,
but only recently have there been concerted attempts at reducing frontal area and drag coefficient simultaneously. The results have been
remarkable.

First, let us look at conventional approaches.
Nonweiler found that mounted cyclists in racing clothes had drag coefficients C_D of about 0.9,
where the average frontal area—of which the bicycle made up an appreciable portion—was
taken to be about 3.6 ft^2 (0.33 m^2).[5] Loose clothing increased the drag area by 30 percent.
There is considerable independent evidence that
0.9 is a reasonable value for the circumstances.
For instance information referred to by Sharp[6]
on the wind resistance experienced by bicyclists
can be interpreted as being based on approximately such a drag coefficient. Wind-tunnel experiments on the upright human form, credited
to A. V. Hill, give about the same value.[7]
Rouse's account of aerodynamic work on the
wind resistance of cylinders[8] can be interpreted
as suggesting that an assembly of short cylinders representing the form a cycle and its rider
would have a drag coefficient of about 1.0. (It
appears unrealistic to quote any value for these

Table 4.1 Values of C_D.

Sports car	0.2–0.3
Sedan	0.4–0.5
Bus	0.6–0.8
Truck	0.8–1.0
Square plate	1.2
Sphere	0.47
Cylinder	0.7–1.3
Streamlined body	0.1
Motorcyclist	1.8
Racing cyclist	0.9
Moped	0.78–1.1

Source: reference 9.

drag coefficients to greater accuracy than the first significant figure because of the magnitude of the experimental errors.)

Drag coefficients for other wheeled vehicles are given in reference 9. The range is from 0.2 for sedan automobiles, to 1.0 for square-ended motor trucks, to 1.8 for a motorcycle and rider. Racing cars have very low drag coefficients, 0.1 or less. Table 4.1 gives these C_D values and table 4.2 gives some estimates for "mopeds" based on published performance data. As would be expected, the moped figures are close to the values for bicycles and riders, but below Kempe's figure for a motorcyclist, which perhaps should be regarded with suspicion.

Table 4.2 Air resistance of mopeds.

Make	Engine power (hp)	Weight of machine, (lb)	Weight of rider (lb)	Max. speed (mph)	Air-resistance data (est.) hp	C_D
Powell	1.05			26	0.64	1.1
Mobylette	1.35	75	200	30	0.86	0.94
Magneet	1.6	115	200	33	1.00	0.78

From the above deliberations emerges a numerical relationship between variables suitable for practical use with everyday units. It is assumed that the vehicles concerned are running at sea level, so that "standard" air density can be assumed. Then, from the definition of the drag coefficient, the following relation can be derived:

$$\text{Drag force (lbf)} = 0.00256 \times C_D \times \text{Frontal area (ft}^2) \times [\text{Speed (mph)}]^2. \quad (4.5)$$

If a bicyclist has a C_D of 0.9, this takes the form

$$0.0023 \times \text{Frontal area (ft}^2) \times [\text{Speed (mph)}]^2. \quad (4.6)$$

In S.I. units, this is

Drag force (newtons) = 0.54 × Frontal area (m²)
                      × [Speed (m/sec)]².

(4.7)

Reducing frontal area

Effect of riding position (conventional bicycles)
In this book, whenever a typical example of a crouched racing bicyclist has been under discussion it has been assumed on the basis of evidence presented by Nonweiler (ref. 5) that the frontal area presented to the wind measures about 0.33 m². For a tourist-type bicyclist (see table 2.2), it has been assumed that the frontal area is about 0.5 m² (these figures were used to calculate curves A and B of figure 2.2). The evidence for the 0.5 m² value is from reference 6 and from experiments by F.R.W. The frontal area is obviously affected by the rider's size, clothing, and position and by the bicycle and the accessories used.

The wind resistance of skiers is relevant. Experimental work[10] has shown, for instance, that the position of the arms is of importance; in the elbows-out position appreciable extra resistance is experienced. The nearest approach of the skiing subject to the position of a typical track bicyclist seems to be the crouch shown in figure 4.4 (bottom). The resistance experienced at 50 mph (22.3 m/sec) was 20.5 lbf (91.3 N). One could reasonably assume that the frontal area of the skier with accessories was near that of a crouched bicyclist and the machine. The drag force can be calculated as before:

0.0023 × 3.1 × 50² = 17.8 lbf (79.2 N).

The fairly close agreement of the estimate and the reported results is satisfying evidence that the data quoted in the foregoing discussion are realistic.

Figure 4.4
Aerodynamic drag of the human body. Four positions demonstrated by skier Dave Jacobs were photographed in a wind tunnel at the same moment the drag was recorded. The air speed was a steady 80 km/h. With erect position (run 9), drag was 216 N. In a high but compact crouch (run 15), drag was reduced by more than half to 91 N. From reference 10.

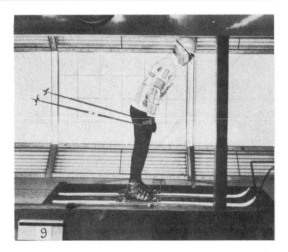

Small-wheeled tricycles

It has often been proposed that a tricycle with smaller-than-usual rear wheels could be faster than a conventional one. If 16-inch (406-mm) wheels could be used on a tricycle, the decrease in frontal area would be about 0.14 ft^2 (0.013 m^2)—a small decrease in comparison with the average total area of rider and machine, which is about 4.1 ft^2 (0.381 m^2). The area would actually be reduced to about 0.96 of the original. The extra 4 percent power should therefore result in a speed increase of 1.3 percent ($1.04^{1/3}$ is about 1.013). It could well be that some of this increase in speed due to lowered wind resistance would be lost because of the greater rolling resistance of small wheels, although the stiffer wheels might counteract this in other ways. In any case, the possible speed increase is very small and there appear to be no grounds for expecting a small-wheeled tricycle to be faster.

Recumbent bicycles

Because most of the area presented normal to the wind by a conventional bicycle-plus-rider is that of the rider, the only method of achieving a marked area reduction is to change the rider's posture. At different times over the last century designers have produced vehicles on which the rider reclines to a small or a great extent; these have led to the so-called reclining or recumbent bicycle.

The most famous of earlier "recumbents" was the Velocar (figure 1.25), introduced in the 1930s, which reduced the frontal area by almost 20 percent. With it a relatively unknown rider broke most existing short-distance track records and beat the reigning world champion. The governing body of cycle racing outlawed the recumbent bicycle before it could do any more damage to existing reputations.

In 1974 a series of speed trials and races with very few design restrictions was started by

Chester Kyle in Long Beach, California. The governing organization became the International Human-Powered Vehicle Association, which seeks to spread out into air- and water-vehicle competitions. Technical and athletic progress under these free design rules has been very impressive. Most early entries were conventional machines with streamlined enclosures, but recently all winning entries have been streamlined recumbents. Let us consider developments in that order.

Figure 4.5
Some past attempts at streamlining bicycles. Courtesy of *Cycling*.

Reducing drag coefficient by streamlining

Streamlined enclosures are not allowed in bicycle racing under the rules of the International Cycling Federation, the body that also outlawed recumbent bicycles. However, for special events, complete streamlined casings have been made. Some past attempts at streamlining are shown in figure 4.5. On average, these enabled the riders' top speeds to increase from about 30 to about 36 mph. From these figures one can calculate that the drag coefficients of the casings were about 0.25. The more recent intensive work by Kyle and competitors in the IHPVA trials has led to enclosures with much lower drag coefficients, in the region of 0.1. Table 4.3, taken from Kyle and co-workers,[11] shows drag coefficients for full and partial fairings. The Kyle fairing with a measured drag coefficient of 0.10 was based on NACA wing profile 0020 (ref. 1). This low drag coefficient was achieved despite a ground clearance of 150 mm. Earlier tests by Kyle and co-workers[12] had seemed to indicate that a small, almost rubbing, ground clearance was helpful in reducing drag. The importance of also using a length-to-thickness "fineness" ratio of 4 or more to avoid boundary-layer separation and a large drag increase was emphasized in reference 13.

It is noteworthy in table 4.3 that worthwhile drag reductions were given by partial fairings such as those shown in figures 4.6 and 4.7. (In some cases, these were simply curved sheets mounted on the handlebars.) The principal reason for this kind of drag reduction appears to be the reduction in the effective area. Without any fairing, the air billows out around the rider's bluff body, disturbing air over an area much greater than that of the body alone. An upstream streamlined shape (such as a partial fairing) can reduce this disturbance. A partial fairing can also have a favorable distribution of pressure. The effective drag coefficient for a front fairing, such as the lowest one in figure 4.8, can actually be negative (ref. 2, p. 3.12).

Table 4.3 Drag and speed characteristics of streamlined human-powered vehicles.

	Frontal area (m²)	C_D	Drag reduction at 8.9 m/sec	Watts required at 8.9 m/sec	Speed with no power increase (m/sec)	Maximum measured speed (m/sec)
Racing bicycles						
Bare Bicycle	0.50	0.78	0	203	8.89	—
Bicycle + Glen Brown Zipper 2	0.50	0.60	13%	177	9.31	15.20
Bicycle + modified Zipper 2	0.55	0.52	22%	159	9.64	—
Van Valkenburgh Aeroshell	0.65	0.32	34%	125	10.22	15.06
Aeroshell + bottom skirt	0.68	0.21	48%	97	11.06	20.79
Kyle fairing	0.71	0.10	67%	68	2.78	20.77
Recumbents						
Palombo supine tricycle, bare	0.35	0.77	20%	151	9.58	16.17
Palombo tricycle with fairing	0.46	0.28	52%	92	11.36	19.84
Van Valkenburgh prone quadracycle with fairing	0.46	0.14	68%	64	13.03	22.08

Source: reference 11.

Figure 4.6
Campbell fairing.
Courtesy of Jack
Campbell.

Figure 4.7
Windfoiler fairing.

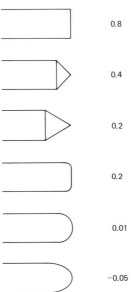

0.8

0.4

0.2

0.2

0.01

−0.05

Figure 4.8
Partial drag coefficients
of front fairings. From
reference 2.

Streamlining the tubing could reduce the wind resistance of the bicycle itself by half at high speeds. Nonweiler suggests that the bicycle's resistance could amount to about 1/3.6 of the total wind resistance (ref. 5). If streamlining the tubes reduced the wind resistance by half, the effect on the total wind resistance (machine plus rider) would be $1/(3.6 \times 2)$, or 1/7.2. A conservative view would be to take the reduction as 10 percent from the original wind resistance. In the late 1970s, some companies began producing "aerodynamic" frames, crank sets, and handlebars. A possible danger in this is that an oval handlebar tube, for instance, might be much more susceptible to fatigue. However, in a test of a standard track bicycle against one with streamlined components, the latter was 2 percent faster over a flying kilometer.[14]

At racing speeds, the power to propel rider and machine is almost all spent in overcoming air resistance, and this power is proportional to the speed cubed. If, therefore, the wind resistance is reduced by $\frac{1}{10}$, the speed will have increased, for the same power, by approximately the cube root of $(1/\frac{1}{10})$. This ratio of speeds is 1.03, so there is a 3 percent increase in speed. Whether or not the rider thinks this worthwhile is a personal opinion, but records have been broken with much smaller increments.[15]

A French company has started producing bicycles with lower-drag tubes and wheels, and Japanese companies are manufacturing aerodynamically shaped cranks. The effects will be small, but any gain achieved at no real cost is worthwhile.

Combined effects of recumbent posture and streamlining

Winning speeds in the various categories of the IHPVA trials are shown in figure 4.9. In the 1980 200-m flying-start trials, a single-rider machine won at over 56 mph (25.3 m/sec); the multiple-rider category was won at almost 63 mph (28.1 m/sec). These performances seemed out of reach in 1978. All recent winners have

Figure 4.9
Winning speeds in
IHPVA races. Top two
lines are for 200-meter
flying start; solid line is
for multiple-rider
vehicles (with number of
riders given in
parentheses) and dashed
line is for single rider
vehicles. Dash-dot line is
for one-hour standing-
start single-rider-vehicle
race. Adapted from
Human Power, the
IHPVA newsletter.

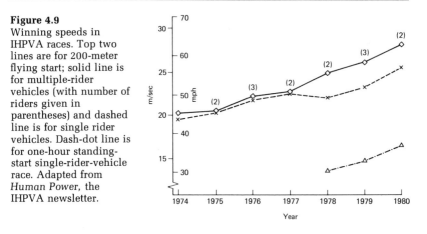

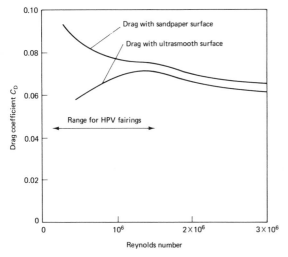

Figure 4.10
Drag coefficient as a
function of Reynolds
number for a low-drag
fairing of circular cross-
section. Reynolds
number = (Relative
velocity × Maximum
diameter of fairing) /
Air viscosity. From
reference 18.

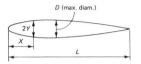

Figure 4.11
Definitions of fairing dimensions used in table 4.4.

Table 4.4 Geometrical configurations for low-drag fairings (see figure 4.11) derived via the formula $Y^2 = a_1X + a_2X^2 + a_3X^3 + a_4X^4 + a_5X^5 + a_6X^6$, where $a_1 = +1.000000$, $a_2 = +0.837153$, $a_3 = -8.585996$, $a_4 = +14.075954$, $a_5 = -10.542535$, and $a_6 = +3.215422$. Length/Diameter = 7.00, Nose radius/Maximum diameter = 0.714, and Tail radius/ Maximum diameter = 0.0143.

X/L	Y/D	X/L	Y/D
0.00	0.0000	0.84	0.2652
0.02	0.1423	0.86	0.2429
0.04	0.2020	0.88	0.2193
0.06	0.2476	0.90	0.1941
0.08	0.2855	0.92	0.1672
0.10	0.3179	0.94	0.1383
0.12	0.3462	0.96	0.1065
0.14	0.3710	0.98	0.0699
0.16	0.3930	1.00	0.0000
0.18	0.4123		
0.20	0.4260		
0.22	0.4439		
0.24	0.4565		
0.26	0.4674		
0.28	0.4765		
0.30	0.4841		
0.32	0.4900		
0.34	0.4944		
0.36	0.4976		
0.38	0.4994		
0.40	0.5000		
0.42	0.4995		
0.44	0.4978		
0.46	0.4950		
0.48	0.4911		
0.50	0.4864		
0.52	0.4806		
0.54	0.4739		
0.56	0.4665		
0.58	0.4580		
0.60	0.4486		
0.62	0.4384		
0.64	0.4273		
0.66	0.4154		
0.68	0.4026		
0.70	0.3890		
0.72	0.3743		
0.74	0.3588		
0.76	0.3422		
0.78	0.3245		
0.80	0.3059		
0.82	0.2861		

used fairings with drag coefficients in the region of 0.1 at the maximum speeds and supine or prone riding positions. The transmissions have been conventional crank-and-chain types. In some cases the rear riders in the multiple-rider machines have used their hands as well as their legs for the 200-m event.

Aerodynamic effects of passing vehicles

All bicyclists who have ridden on roads frequented by large, fast-moving motor vehicles have experienced side-wind forces from their passing, but no experimental work concerning the magnitude of the lateral forces on actual bicyclists seems to have been reported. However, Beauvais has reported valuable work concerning wind effects upon "parked" and jacked-up $\frac{1}{10}$-scale model automobiles.[16] (There is considerable concern in the United States about the safety of jacked-up vehicles at the sides of expressways.) From Beauvais's data we can estimate that a bicyclist may experience lateral forces of several pounds when overtaken closely by a large vehicle moving at 70 mph. The laws prohibiting bicycling along expressways are reasonable.

Drafting

"Taking pace" or "drafting" is when a bicyclist travels close behind another moving body, using it to "break the wind." The vortices behind a leading bluff body (see figure 4.1) may indeed help to propel the trailing rider. Drafting is therefore an important part of the strategy in massed-start races. Only recently have quantitative data been taken on the assistance given by drafting.[17]

The second rider ("stoker") of a tandem is drafting behind the leading rider, and therefore incurs little additional drag.

When streamlined fairings are used, competitors soon find that there is no benefit in drafting because there are no trailing vortices or large masses of captured air behind an aerodynamically faired shape.

References

1. I. H. Abbott and A. E. Doenhoff, *Theory of Wing Sections* (New York: Dover, 1959).

2. S. F. Hoerner, *Fluid Dynamic Drag* (Bricktown, N.J.: Hoerner, 1959).

3. R. A. C. Fosberry, Research on the aerodynamics of road vehicles, *New Scientist* 6 (20 August 1959): 223–227.

4. A. Sharp, *CTC Gazette* (January 1899): 11.

5. T. Nonweiler, Air Resistance of Racing Cyclists, report 106, College of Aeronautics, Cranfield, England, 1956.

6. A. Sharp, *Bicycles and Tricycles* (London: Longmans, Green, 1896; Cambridge, Mass.: MIT Press, 1977).

7. G. A. Dean, An analysis of the energy expenditure in level and gradient walking, *Ergonomics* 8 (1965), no. 1: 31–47.

8. H. Rouse, *Elementary Mechanics of Fluids* (London: Chapman and Hall, 1946), p. 247.

9. *Kempe's Engineer's Year Book*, vol. 11 (London: Morgan, 1962), p. 315.

10. A. E. Raine, Aerodynamics of skiing, *Science Journal* 6 (1970), no. 3: 26–30.

11. C. R. Kyle, V. J. Caizzo, and P. Palombo, Predicting human-powered-vehicle performance using ergometry and aerodynamic-drag measurements, Proceedings of IMFA Conference, Technical University of Cologne, 1978.

12. C. R. Kyle, The aerodynamics of man-powered land vehicles, Third National Seminar on Planning, Design, and Implementation of Bicycle and Pedestrian Facilities, San Diego, Calif., 1974.

13. C. R. Kyle and W. E. Edelman, Man-powered vehicle design criteria, Third International Conference on Vehicle-System Dynamics, Blacksburg, Va., 1974.

14. K. Evans, Aerodynamics the keynote, *Cycling* (11 October 1980): 14–15.

15. F. R. Whitt, Is streamlining worthwhile?, *Bicycling* (July 1972): 50–51.

16. F. N. Beauvais, Transient aerodynamical effects on a parked vehicle caused by a passing bus, Proceedings of First Symposium on Road Vehicles, City University of London, 1969.

17. C. R. Kyle, Reduction of wind resistance and power output of racing cyclists and runners travelling in groups, *Ergonomics* 22 (1979), no. 4: 387–397.

18. M. Gertler, Resistance Experiments on a Systematic Series of Bodies of Revolution, for Application to the Design of High-Speed Submarines, report C-297, David Taylor Model Basin, U.S. Navy (1950).

Recommended reading

A. L. Minter, Optimising of petrol-engine design, *Engineering* (London) 213 (1973), no. 12: 896–900.

C. R. Kyle, How Accessories Affect Bicycle Speed, engineering report 75-I, California State University, Long Beach, 1975.

M. A. Van Baak, The Physiological Load During Walking, Cycling, Running, and Swimming and the Cooper Exercise Program (Keppel: Krips Repro, 1979).

C. R. Kyle, The Aerodynamics of Bicycles, report, California State University, Long Beach, 1979.

5

The wheel

Traveling by foot requires a severalfold range of power for both hard and soft ground, and walking can be said to be a reasonably adaptable means of locomotion. The resistance to motion of a wheel, however, can vary several hundredfold from pavement to soft soil. Hence, there was a real incentive to develop paved roads when wheels were adopted for horse-drawn vehicles (figure 5.1). The Roman empire was the first civilization to make use of this idea. It is recorded that the times taken to travel across various European routes to Rome were shorter in that era than a thousand years later in the Middle Ages, when the Roman road system had vanished through lack of maintenance.

After the Middle Ages, as the stultifying effects of spiritual opposition to technological change were overcome, inventions to improve everyday life appeared rapidly. Among these were iron-covered wooden railways, followed by iron wheels and cast-iron rails (1767). This gave rise to the railway age of Victorian times, and was paralleled by the reappearance of a fair number of paved roads. Thomson (1845) and Dunlop (1888) introduced pneumatic tires, which decreased the rolling resistance of carriage wheels to something closer to that experienced by the railway wheel and also introduced a degree of comfort on common roads. Ever since that time there has been competition between the low-friction guided rolling of vehicles on tracks and the greater freedom of steerable road vehicles with pneumatic tires. It has been established beyond doubt that steel wheels on steel tracks require the least power of all systems used to drive practical vehicles at a given constant speed. The power consumed in rolling the most flexible pneumatic-tired wheel is several times

Figure 5.1
Replica of Egyptian chariot wheel of 1400 B.C. Note rawhide wrapping to make tire resilient. Courtesy of Science Museum, London; reproduced with permission.

greater, and the average automobile wheel on the best surfaces generally available has ten or more times the resistance to motion of a train wheel on its track.

Rolling resistance

The power needed to propel wheeled vehicles depends not only on the ease of rolling of the wheels but also on the physical properties of the surface. A great deal of information is available concerning the former in general and the latter for harder surfaces. (Wheel motion on soft ground is significant mostly to agricultural engineers and designers of military vehicles.)

The term "rolling resistance" as used in this book means the resistance to the steady motion of the wheel caused by power absorption in the contacting surfaces of the wheel and the road, rail, or soil on which it rolls. It does not normally include the bearing friction or the power needed to accelerate or slow the wheel because of its inertia. The energy lost in acceleration is, for bicycle wheels, of small consequence compared with the power absorbed by tire and road; unfortunately, it is often referred to in the sense of "ease of rolling of wheels" and can be twisted into the statement that "little wheels roll more easily than large wheels." This latter is only partly true, even if "rolling" is taken to mean "accelerating and decelerating."

Bicycle wheels are now of such a pattern that design changes can have only small effects on acceleration properties, but a wheel of a given diameter has a rolling resistance (in the sense of surface power absorption) of approximately half that of a wheel of half the diameter. This definition of rolling resistance, as accepted in the engineering literature, implies that the weights of rider and machine, both greatly exceeding that of the wheels, influence the motion of the bicycle via the tires; multiplying the coefficient of rolling friction (rolling resistance divided by vertical load applied to wheel) by the weight gives the rolling resistance.

Train wheels

The rolling of railroad wheels has been investigated thoroughly.[1] It is more amenable to measurement than other wheel-rolling actions, such as that of pneumatic-tired wheels on roads. The hardness of the railway wheel and the track can be specified closely and are less variable than other types of contacting surfaces.

The wheel's rolling resistance is caused by the deformation of wheel and track, which produces a temporary "dent" or "sinkage" (figure 5.2). This deformation causes the point of instantaneous rolling of the wheel to be always ahead of the point geometrically directly below the wheel's center of rotation about its bearing, which is attached to the vehicle. The deformation of the wheel, or of the wheel and the track, results in the reaction forces shown in figure 5.2. These forces are higher in the "compression" stage than during the subsequent expansion because of hysteresis (the internal friction of materials). Thus, a pair of forces exerting a retarding torque (known as a "couple") is set up. The numerical value of the torque is the downward force between wheel and surface, which in steady state is the weight of the wheel, plus its share of the weight of the vehicle, multiplied by the distance $b/8$, where b is the length of the wheel-to-ground contact.

Reference 1 shows why the displacement of the instantaneous center of rotation can be calculated as the length b divided by 8. Experiments with railway wheels of typical diameters resting on rails have shown that the distance $b/8$ can be taken as 0.01–0.02 inches (0.25–0.50 mm). It is thus possible to calculate the rolling resistance according to the method of reference 1. If the wheel's radius is 20 inches (508 mm), the calculated coefficient of rolling friction ranges from 0.0005 to 0.001, in addition to bearing resistance.

One can check the calculation in the preceding paragraph using information given on page

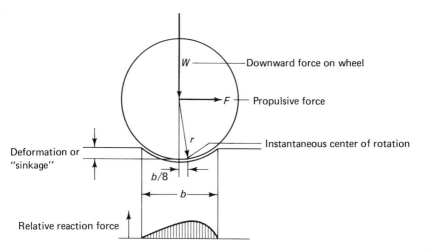

W ——————— Downward force on wheel

F —— Propulsive force

Instantaneous center of rotation

Deformation or "sinkage"

r

b/8

b

Relative reaction force

Figure 5.2
Resistance diagram of
rolling wheel.

F532 of the *Engineering Encyclopedia* (New
York: Industrial Press, 1954). This source is unusual in that it gives a quantitative relationship
for the coefficient of rolling friction, C_R, of cylinders on plane surfaces, including the effect of
cylinder diameter:

$$C_R = \frac{f}{\text{Diameter in meters}}.$$

The experimentally determined values of f
quoted include a range of 0.0002–0.0060 for
steel-on-steel contact. For a wheel or a pair of
wheels 686 mm in diameter, loaded with 80 kg,
the resistance to rolling is

$$\frac{80 \times 9.81 \times 0.0012}{0.686},$$

which yields 1.4 N (0.3 lbf).

Wheels on soft ground
The general effect of wheel form on rolling resistance was investigated over a century ago by
Grandvoinet,[2] who found that if the diameter of
the wheel was increased 35 percent the rolling
resistance on soft ground decreased 20 percent.
A similar increase in width decreased the roll-

ing resistance by only 10 percent. For a very large wheel, it has been found that the tread width has a negligible effect on rolling resistance.[3] Other studies investigated the once-common steel-rimmed wooden agricultural wheel. The characteristics of modern pneumatic-tired military and agricultural vehicles are still being investigated. Not all concerned subscribe to the theory that these large-tired wheel vehicles can "float" on soil, as might be thought.

In passing, it is worth noting that driving a wheel on soft ground may require more effort than walking or running, which, whether associated with man or quadruped, are mechanisms of a different character. Races between bicyclists and runners over rough country show that the speeds of the two are much closer than in races on hard ground.

A great deal of experimental work has been carried out in more recent times on the power needed to move agricultural vehicles. Barger et al. (ref. 3) verified the general effects of wheel cross-sectional shape and diameter as postulated by the very early workers, and also carried out investigations on pneumatic tires. The main findings have been that wheel diameter, whether for a steel-rimmed wheel or for a pneumatic-tired one, is an important factor. The larger the wheel the more easily it runs when supporting a given weight, whether the surface is soft or hard. For hard ground the ease of running can be related to the diameter by a simple inverse-proportion formula; for soft ground the effect of diameter is even greater.

When a loaded pneumatic or steel tire presses on a road surface, the shape of the area of deformation of the surfaces is much influenced by, among other things, the diameter of the wheel. Taking account of the relative dimensions of the contact areas, and reasoning along the lines employed for the railway-train wheel (see figure 5.2), one can deduce that the forces opposing rolling are inversely proportional to the wheel diameter. (Readers interested in rolling-friction theory are advised to consult references 4 through 7 for further details about a subject not frequently referred to in textbooks on basic physics.)

Pneumatic tires and their properties

The pneumatic-tired wheel rolling on the road exhibits exaggerated characteristics in comparison with the steel wheel on rails. For instance, the flattening of the tire over an "equivalent" distance b (see figure 5.2) is obviously much greater for pneumatic tires, and therefore the theory predicts a much greater rolling resistance, as found in practice. What is very difficult to predict is the effect of flexing of the tire walls, which is so dependent upon inflation pressure and the design of the carcass, as compared with the constancy of steel's elasticity. An interesting peculiarity of pneumatic tires is that they affect steering properties, because any side force applied to the axle is resisted by the road at a point on the tire that is not directly beneath the axis[8,9] but slightly behind. This results in a measurable "twisting effect" not experienced by hard wheels on hard surfaces. This is called *self-aligning torque,* and is a measure of the tendency of the steered wheel to follow the direction of motion. Tire-inflation pressure and carcass flexibility obviously also influence this twisting effect, as they do rolling resistance.

Early bicycles had solid rubber tires. The record times for the mile track for both the solid-rubber-tired "old ordinary" and the solid-rubber-tired "safety" are almost the same, close to $2\frac{1}{2}$ minutes. It is known that the high bicycle offers greater wind resistance and requires more riding skill than the smaller-wheeled bicycle. Hence, the above findings support the explanation that the bigger wheel runs more easily than the smaller wheel (see ref. 5)—the lower rolling resistance compensates for the higher wind resistance.

Although it might at first sight appear that there are too many factors influencing pneumatic-tire rolling for any simple correlation to be devised, in practice this is not so. The predominant variables have been found to be tire-inflation pressure, wheel diameter, and road surface. Actual road speed has an effect, but not until speeds well above those common for bicycles are involved is it appreciable.[10] For modern bicycles running on hard roads, the range of each of the three predominant variables is only about twofold, giving a total possible effect of some eightfold on the rolling resistance.

The shapes and dimensions of the marks made by bicycle tires upon the road were investigated by the senior author using carbon paper and thin white paper. Tire loads were kept constant at 90 lbf (400 N), and pressings were made with various inflation pressures (15–75 lbf/in.^{2}; 103–517 kPa, where 1 pascal $\equiv$ 1 newton per square meter) and various tire sizes (12 $\times$ 2, 16 $\times$ 1$\frac{3}{8}$, and 27 $\times$ 1$\frac{1}{4}$ in.; 305 $\times$ 51, 406 $\times$ 44, and 686 $\times$ 32 mm). (The sizes given here are direct conversions of the nominal "inch" sizes, and do not necessarily give the actual outside diameter.) The maximum lengths of the impression were compared with experimentally determined rolling-resistance coefficients in two ways: in one case the length was used along with the outside diameter of the inflated tire; in the other it was used to estimate the "sinkage" of the inflated tire under load. The results are shown in figures 5.3 and 5.4. On the same figures we have plotted data from references 11–13 concerning pneumatic 28 $\times$ 6 and 43 $\times$ 7$\frac{1}{2}$ inch (711 $\times$ 152 and 1,092 $\times$ 190 mm) tractor tires with inflation pressures of 10–40 lbf/in.^{2} (69–276 kPa). Figure 5.5 shows imprints and contact prints of various tires and wheels.

From the mean lines drawn through the experimental points in figures 5.2 and 5.4 we can derive the following simple approximate relationships for pneumatic tires running on

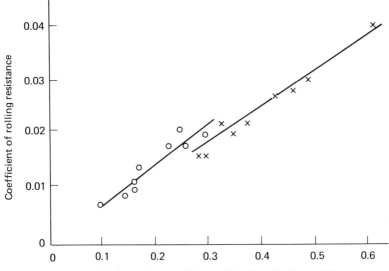

Figure 5.3
Relationship between coefficient of rolling resistance and length of tire marking on road combined with tire outside diameter (under static conditions), for small tire deformation or sinkage, where (Length of tire marking)² = 4 × Tire diameter × Sinkage. (○) Bicycle tires, (×) tractor tires.

hard concrete surfaces, where C_R is the coefficient of rolling resistance:

$$C_R = \text{const} \times \frac{\text{Sinkage}}{\text{Max. length of road impression}}$$

$$(5.1)$$

when the constants are 0.33 (standard load 90 lbf [400 N]) for bicycle tires and 0.25 (standard load 1,000 lbf [4.45 kN]) for tractor tires, and

$$C_R = \text{const} \times \frac{\text{Max. length of impression}}{\text{Diameter of inflated tire}} \quad (5.2)$$

when the constants are 0.070 for bicycle tires and 0.064 for tractor tires.

Sabey and Lupton measured the markings of static and moving motorcar tires.[14] There is some decrease in length of tire marks due to movement, but it can be deduced that at the low speeds common in bicycling the statically measured lengths are only a few percent greater than the moving lengths.

Figure 5.4
Relationship between
coefficient of rolling
resistance and length of
tire marking on road
combined with sinkage
caused by load on tire
under static conditions.
($\triangle$) Bicycle tires; ($\times$)
tractor tires.

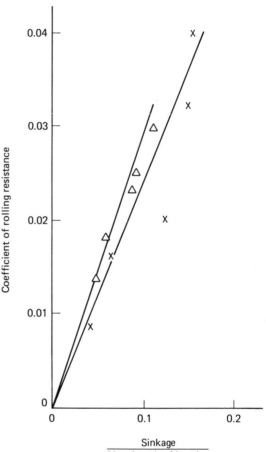

As stated above, the rolling resistance of pneumatic tires is a combination of several resistances, not all of which can be predicted theoretically. Thus, experiments are needed (nowadays, most of these involve towed wheels[15]). The data yielded by these experiments are discussed further below.

Formulas for calculating the rolling resistance of automobile tires about 5 inches in cross-section are given in references 2 and 16, and some

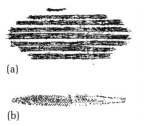

(a)

(b)

(c)

Figure 5.5
(a) Imprint of $12\frac{1}{2} \times 2\frac{1}{4}$-inch bicycle tire inflated to 26 1bf/in.^{2} with 90-lbf load (actual length of impression: 4 inches).
(b) Imprint of $27 \times 1\frac{1}{4}$-inch bicycle tire inflated to 40 lbf/in.^{2} with 90-lbf load (actual length 3.8 inches).
(c) Contact prints of 35-inch-diameter steel train wheel on steel rail with load of 6,075 lbf (actual lengths up to $\frac{3}{4}$ inch). From F. R. Whitt, Tyre and road contact, *Cycle Touring* (U.K.) (February–March 1977): 61.

information concerning bicycle tires $1\frac{1}{4}$–2 inches in diameter is given in references 17 and 18. All information indicates that the most important factor in ease of rolling is the inflation pressure (if road surface, wheel size, and cross-section are similar). It seems probable that the C_R of a bicycle tire on a 26-inch wheel, on smooth roads, ranges from 0.010 to about 0.0054 if the inflation pressure is varied from 17 lbf/in.^{2} (at which the rim is liable to bump on the road and give warning of gross misuse to the careless rider) to the 75 lbf/in.^{2} recommended by tire makers (table 5.1). Less smooth or hard surfaces, such as rough macadam or gravel, may cause an increase of 50–100 percent. For a given surface roughness and a given load, the larger the wheel the easier it rolls—a fact also established over centuries by experience with horse-drawn vehicles.

The earliest accessible information on the bicycle tire seems to be that given by Sharp in reference 18 (see table 5.2). He quotes three values for the coefficient of friction of tires on road and track from a publication by C. Bourlet. No tire pressures are specified, although it was well known by then that pressure had a major influence on rolling ease. Some more recent experiments by Patterson (ref. 17) are summarized in table 5.3.

Available data on the effect of tire pressure and wheel diameter on rolling resistance are combined in figure 5.6. Because no tire pressures are quoted in the material Sharp credits to Bourlet, it has been necessary to assume that appropriate limits are 55–80 lbf/in.^{2}. The two formulas quoted below predict similar values for C_R and little effect from vehicle speed in the low range of speed, applicable to cycling, of up to about 12 mph. (If the curves had been calculated for 30 mph, the C_R values would have been only a few percent higher.) These formulas and others were discussed at length by Ogorkiewicz (ref. 10), who also stressed the applica-

Table 5.1 Recommended tire pressures, in lbf/in.² (kPa), for 26- or 27-inch wheels.

Tire cross-section	Tire type	Maximum weight of rider					
		84 lb (38 kg)		140 lb (63.6 kg)		196 lb (89 kg)	
		Front	Rear	Front	Rear	Front	Rear
1¼ in. (31.75 mm)	Regular clincher	30 (206)	45 (310)	40 (276)	55 (379)	50 (345)	65 (448)
	High-pressure clincher	45 (310)	60 (414)	55 (379)	70 (483)	65 (448)	80 (552)
	High-pressure tubular	60 (414)	70 (483)	75 (517)	85 (586)	85 (586)	100 (689)
1⅜ in. (34.9 mm)	Regular clincher	25 (172)	35 (241)	35 (241)	50 (345)	45 (310)	60 (414)
	High-pressure clincher	40 (276)	50 (345)	50 (345)	65 (448)	60 (414)	75 (517)
1½ in. (38.1 mm)	Regular clincher	25 (172)	30 (206)	30 (206)	45 (310)	40 (276)	55 (379)
1¾ in. (44.45 mm)	Regular clincher	20 (138)	25 (172)	40 (276)	35 (241)	50 (345)	
2 in. (50.80 mm)	Regular clincher	20 (138)	25 (172)	20 (138)	30 (206)	45 (310)	

Note: These pressures are recommended in order to avoid too much flexing of the sidewalls of the tires and, especially, bumping of the wheel rim on the road. They do not necessarily give minimum rolling resistance.

Table 5.2 Rolling resistances of early tires.

	Coefficient of rolling friction, C_R		
Suface	Solid tire	Pneumatic tire	Speed (m/sec)
Racing track		0.004	
road[a]		0.005–0.010	
Smooth macadam road[b]	0.022–0.027	0.013–0.016	
Flag pavement[c]	0.027	0.015	2.23
Flint[c]	0.027	0.014–0.017	1.79–4.47
Flag pavement[d]		0.017	
Macadam road[d]		0.016	
Broken granite[d]		0.023	

a. Cycle tires; data from reference 18, p. 251.
b. Car tires; data from A. W. Judge, *The Mechanism of the Car,* vol. III (London: Chapman and Hall, 1925), p. 150.
c. Heavy cycle tires; data from reference 18, p. 256. The high figures quoted here are probably due to the inclusion of air resistance in addition to rolling resistance.
d. Thomson's early pneumatic tires (figure 1.22) on a horse-drawn carriage; data quoted in *Mechanics Magazine,* vol. 50 (1848).

Table 5.3 Experimentally determined tire rolling resistances.

Cross-section		Load		Speed		Inflation		Rolling resistance		
(in.)	(mm)	(lbf)	(N)	(mph)	(m/sec)	(lbf/in.^{2})	(kPa)	(hp)	(W)	C_R
2	50.8	120	534	20	8.94	10	68.9	0.1	74.6	0.016
2	50.8	120	534	20	8.94	18	124.1	0.07	52.2	0.011
2	50.8	120	534	20	8.94	30	206.8	0.05	37.3	0.008
2	50.8	150	667	20	8.94	18	124.1	0.1	74.6	0.013
2	50.8	180	801	20	8.94	18	124.1	0.12	89.5	0.013
$1\frac{3}{4}$	44.4	120	534	15	6.7	18	124.1	0.05	37.3	0.010
$1\frac{1}{4}$	31.7	120	534	15	6.7	45	310.3	0.02	14.9	0.004

Source: reference 17, pp. 428–429.

Figure 5.6
Effect of tire inflation
pressure on rolling
resistance. Curve A: bias-
ply auto tire on smooth
hard surface (data from
reference 16). Curve B:
bias-ply auto tire on
smooth hard surface
(data from reference 4).
Limits C: 28 × 1½-inch
bicycle (average) bicycle
tire on road and track.

Curve D: 27 × 1¼-inch
bicycle tire on smooth
hard surface (from
F.R.W.'s low-speed
experimental data). Curve
E: 16 × 1⅜-inch bicycle
tire on medium-rough
hard surface (from
F.R.W.'s low-speed
experimental data). Curve
F: 27 × 1¼-inch bicycle
tire on medium-rough
hard surface (from

F.R.W.'s low-speed
experimental data).
Points ●: assumed to be
for 26 × 1¼-inch bicycle
tire on steel rollers (data
from reference 17; see
table 5.3). Point ○: 6 ×
1-inch roller-skate wheels
on smooth hard surface
(from F.R.W.'s
experimental data).

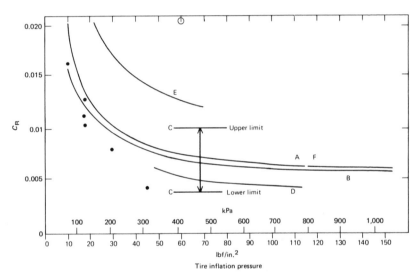

bility of data from curve A of figure 5.6 and
other predictions from the formula—even to
present-day automobile design—although the
basic experimental work was carried out almost
forty years ago, when tires were constructed dif-
ferently than at present. It is probable that the
wheel diameter used was close to that of modern bicycles, 26–27 inches.

With the help of several other cyclists riding
several different bicycles and tricycles on typical roads frequented by bicyclists, F.R.W. has
carried out experimental work on the rolling resistance of tires.[19] All tires used were 1¼ or 1⅜

inch in cross-section and of light construction.
The total weight of rider and machine was always near 180 lb. The experiments showed that,
on concrete or rolled-gravel surfaces, the rolling
resistances were very close to those predicted
by curve A of figure 5.6. This means that light
bicycle tires on rough surfaces have lower resistance coefficients than the larger-cross-section
automobile tires—in other words, bicycle tires
do not require as good a road surface for a given
performance. The results quoted in reference 17
also show that bicycle tires roll more easily than
car tires. Information, in general, suggests that
the performance predicted by curve D can be attained on first-class hard roads with 1¼-inch
light bicycle tires.

Experiments with small-wheeled bicycles
showed that, as Barger et al. (ref. 3) predicted
from their work with pneumatic-tired tractors,
rolling resistance increases in near proportion
as wheel diameter is decreased for a given constant inflation pressure. The small-diameter,
"low-pressure," large-cross-section tire is the
slowest both because of the small diameter and
because of the low design pressure (35 lbf/in.²).

Other data[20] show that the coefficient of rolling
resistance of automobile tires decreases mark-
edly (for example, by 33 percent) during the
first 40 miles of a run. It is not known whether
the same distance-traveled effect applies to bicycle tires.

For comparison, figures 5.7 and 5.8 are included to show how little speed affects the rolling resistance of car tires, although tire-pressure
effects are appreciable from 30 to 50 mph.

Kyle and his associates found a somewhat
larger component of speed in some of their tests
of the rolling resistance of various tires.[21] They
gave the coefficient of rolling friction as a base
value C_{R0} and a speed-dependent increment C_{RV}
so that

$$C_R = C_{R0} + C_{RV} \times V.$$

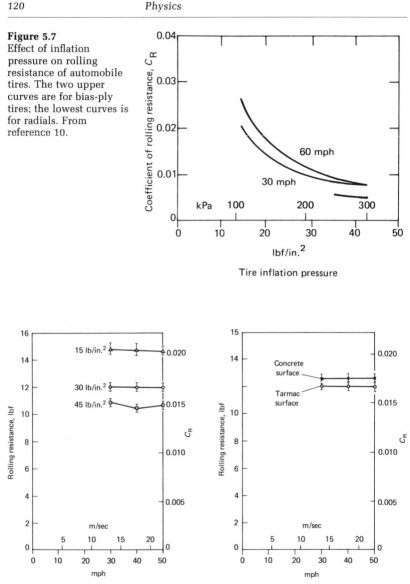

Figure 5.7
Effect of inflation pressure on rolling resistance of automobile tires. The two upper curves are for bias-ply tires; the lowest curves is for radials. From reference 10.

Figure 5.8
Effect of speed on rolling resistance of 5.50 × 16 automobile tire under 720-lbf load. Each point is the mean of six measurements; standard deviations are indicated. (a) Variation of rolling resistance with tire pressure and speed on Tarmac surface.

(b) Variation of rolling resistance with road surface; pressure 30 lbf/in.² (207 kPa). From reference 15.

Their results are shown in table 5.4.

Table 5.5 is included to show how great is the rolling resistance of steel-tired wheels on roads compared with that of pneumatic tires inflated to high pressure. No doubt this fact was immediately apparent to riders of the early "boneshakers." Those machines, in their latter days, were often manufactured with rubber tiring attached to their wheels in a manner adopted many years afterward by makers of horse-drawn carriages. Hollow, square-section rubber tiring was also used, as well as solid tiring, even as early as 1870.

Curve D of figure 5.5 and information on wind resistance given in chapter 4 have been used to compile tables 5.6 and 5.7, which show how tire pressures affect the speed of a bicyclist under various conditions. In particular, note the prediction that a tricycle's extra wheel and axle will make it 5–10 percent slower, for a given power, than a comparable bicycle. This is substantiated by record times for the two types of machine. The effect of good solid-rubber tires is also revealed in table 5.7, which indicates that their coefficient of rolling resistance—0.015—is about the same as that of a pneumatic tire inflated to about 12 lbf/in.^{2} (see table 5.1). This should be interesting to riders of old bicycles and tricycles, who are certainly aware of the slowing effect of solid-rubber tires.

The power needed to overcome rolling resistance is given by

hp = Coefficient of rolling friction
    $\times$ Vertical load (lbf) $\times$ mph/375,

or, in S.I. units,

W = Coefficient of rolling friction
    $\times$ Vertical load (newtons) $\times$ m/sec.

The vertical load of a mass or weight, in newtons, is the product of the mass (kg) and the gravitational acceleration g (m/sec^2). At sea level, g = 9.806 m/sec^2.

Table 5.4 Rolling resistances of various tires.

	Pressure		Vehicle	C_{R0}	C_{RV} [(m/sec)^{i1}]
	lbf/in.^{2}	kPa			
Vittoria imperforable Seta 27-in. tubular	105	724	Bicycle	0.0029	0.165×10^{83}
Criterium 250 27-in. tubular	105	724	Bicycle	0.0039	0.172×10^{l3}
Clement Criterium Seta Extra 27-in. tubular	105	724	Tricycle	0.0019	Very small
Hutcheson 27 $\times$ 1⅛-in. clincher	60	414	Bicycle	0.0047	0.291×10^{u3}
Hercules 26 $\times$ 1⅜-in. clincher	40	276	Bicycle	0.0066	?
United 21 $\times$ 2¼-in. clincher	40	276	Tricycle	0.0061	?

Derived from reference 21.

Note: The surfaces used for these tests were very smooth—similar to those used for cycle tracks, and much smoother than those common road surfaces listed in table 5.5.

Table 5.5 Rolling-resistance coefficients of four-wheeled wagon (steel tires) and 1½-ton stagecoach.

Surface	C_R[a]	Speed	Vehicle
Cubical blocks	0.014–0.022	Slow	Wagon
Macadam	0.028–0.033	Slow	Wagon
Planks	0.013–0.022	Slow	Wagon
Gravel	0.062	Slow	Wagon
"A fine road"	0.034–0.041	4–10 mph	Stagecoach
Common earth road	0.089–0.134	Slow	Wagon

Source: reference 4, p. 683.

Note: The *Engineering Encyclopedia* (New York: Industrial, 1054) gives values in a similar range for steel-rimmed wheels 24–60 inches in diameter.

a. Coefficient of rolling friction.

Table 5.6 Total rolling resistance of pneumatic-tired bicycle wheels calculated for 170-lbf load (figure 5.5, curve D).

Speed (mph)	Pressure[a] (lbf/in.²)	Rolling resistance (hp)	C_R[b]	Total hp lost[c]	Wheel diam. (in.)	Percentage speed reduction for given power due to small wheel
30	75	0.070	0.005	0.69	27	
30	17	0.140	0.010	0.767	27	
29.3	75	0.113	0.008	0.69	16	2.3
12.5	75	0.029	0.005	0.074	27	
12.5	17	0.058	0.010	0.103	27	
11.4	75	0.044	0.008	0.074	16	8.8
5	75	0.0116	0.005	0.0140	27	
5	17	0.0233	0.010	0.0265	27	
3.6	75	0.0138	0.008	0.014	16	28
9.8	35	0.0337	0.008	0.053	16	

a. For 1¼-inch tire.
b. Coefficient of rolling resistance.
c. Including air resistance.

Table 5.7 Effect of tire pressure on propulsive power needed.

Vehicle	Load (lbf)	Speed (mph)	Pressure[a] (lbf/in.²)	Rolling resistance (hp)	C_R[b]	Total hp lost[c]	Percent increase to total power[d]
Bicycle	170	25	75	0.059	0.0051	0.407	
Bicycle	170	25	17	0.118	0.0103	0.466	14
Bicycle	170	12.5	75	0.0295	0.0051	0.074	
Bicycle	170	12.5	17	0.059	0.0013	0.103	39
Tricycle	180	23.5	75	0.082	0.0077	0.407	
Tricycle	180	23.5	17	0.164	0.0155	0.489	20[e]
Tricycle	180	13.4	75	0.476	0.0077	0.11	
Tricycle	180	13.4	17	0.0952	0.0155	0.157	43
Bicycle	170	25	N.A.	0.154	0.0134	0.508	25[f]
Bicycle	170	12.5	N.A.	0.078	0.0134	0.122	65[f]

a. For 1¼-inch tire.
b. Coefficient of rolling resistance.
c. Including air resistance.
d. For "standard" 27-inch wheel and pressure of 75 lbf/in.².
e. Tricycle is 6% slower than bicycle.
f. For solid tires ⅝ inch in cross-section.

Unlike the power needed to overcome wind resistance, which is proportional to the speed cubed, the power lost in rolling is directly proportional to the speed (at least at low speeds).

It can be estimated from tire formulas that a bicyclist who had only rolling friction to overcome should attain speeds of over 100 mph (44.7 m/sec) on a good surface. World records for bicyclists riding behind motor vehicles indicate that 150 mph (67 m/sec) can be attained over short distances, thus verifying the estimation. (It is arguable that air friction is not merely brought to zero, but may actually help to propel a rider pedaling behind a moving shield.) It is probable that similar shielding from the wind would improve a runner's maximum speed of about 20 mph (8.94 m/sec) only slightly, because air-friction effects for a runner are relatively low compared with the other resistances at this speed.

Other interesting conclusions can be drawn. For instance, at maximum bicycle speeds, if the bicycle had no friction or mass and only its air drag resisted motion the top speed would increase by only a few percent. At low speeds, the situation is rather different: at about 10 mph (4.5 m/sec) such a machine would require about half the power needed from the rider under normal conditions. If the same power were to be exerted on a weightless, frictionless machine, the speed would be increased by about 30 percent, to 13 mph (5.8 m/sec).

Advantages and disadvantages of small-wheeled bicycles

In recent times there have appeared on the market bicycles with wheels 14–20 inches (355–508 mm) in diameter—considerably smaller than the common diameter of 26 or 27 inches. Small wheels appear to be accepted as essential if a bicycle is to be easily stowed in the trunk of a car and if one machine is to be safely ridden by people of different heights. In addition, luggage can be carried more easily over a smaller wheel, simply because there is more space available.

And some designers have incorporated spring-ing into small-wheeled bicycles. It appears that these requirements are considered important for those of the general public who may not care to use a conventional machine.

A question often raised concerns the effect of the smaller wheels on the propulsive power needed. Of course the extent to which this power requirement, for a given speed under specified conditions, exceeds that of a conven-tional machine depends on other details of the particular design as well as on wheel size. Of great importance is the tire inflation pressure at which the machine can be ridden with comfort. "Soft" tires add resistance for all wheel sizes, whether the low pressure is intended by the de-signer or is due to the rider's lack of strength or memory. (A desirable pressure for 26–27-inch-diameter $1\frac{3}{8}$-inch tires is about 55–60 lbf/in.²). The effect of inflation pressure on rolling power for two wheel sizes is illustrated in figure 5.9.

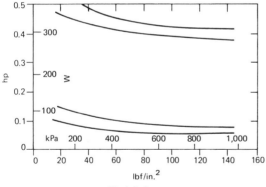

Tire-inflation pressure

Figure 5.9
Effect of tire pressure and wheel diameter on propulsive power required for bicycles. Note diminished rate of decrease of power required at pressure above 75 lbf/in.² (515 kPa), the manufacturer's recommended pressure. Top pair of curves are for 16-inch (upper curve) and 27-inch (lower curve) wheels at 25 mph (11.2 m/sec); bottom pair are for 16-inch (upper) and 27-inch (lower) wheels at 12.5 mph (5.6 m/sec).

Figure 5.10
Slowing effect of 16-inch
(406-mm) wheels
compared with 27-inch
(686-mm) wheels at same
power level. The 27-inch
wheels are assumed to be
running on a smooth
road surface with a
rolling-resistance
coefficient of 0.005, with
a crouching 150-lbm
(68.1-kg) rider presenting
a frontal surface area of
3.65 ft.^{2} (0.34 m^2). The
drag coefficient is 0.9.
The percentage drop in
speed for a "slower"
machine with a rolling-
resistance coefficient of
0.008 and a frontal area
of 5.5 ft^2 (0.34 m^2) is not
very great. Point ● is a
single estimation for
such conditions. In both
cases the tire pressure is
75 lbf/in.^{2} (517 kPa).

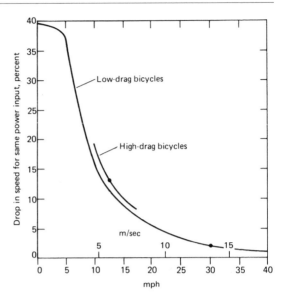

We have estimated the rolling and air resistances of the popular 16-inch (406.4-mm)
wheels and compared the power requirements
at different speeds with those of the 27-inch
wheels (tables 5.6, 5.7; figure 5.10). These calculations show the effect of different pressures
and wheel diameters on the power needed for
riding on very good roads. It is obvious that the
smaller wheels are "slower" over the whole
range of speeds, and to an appreciable extent at
the lower speeds. (If rougher roads had been assumed, the "slowness" would have been more
apparent—unless the wheels had been assumed
to be incorporated in a sprung, damped suspension, in which case they can be superior.) At
25–30 mph and higher, the effect of smaller
wheels is relatively small, according to the calculations, because wind effects are predominant. This accounts for the fact that racing times
of 27-inch-wheeled machines are closely approached by those of smaller-wheeled machines.
Whether the increased resistance of the smaller

wheels at utility and touring speeds of 10–12 mph is acceptable depends on the temperament of the rider.

The rolling resistance R may be calculated by the methods of references 21 and 16:

$R = C_R W$,

where W is the weight of machine plus rider and C_R, the coefficient of friction, is given by

$$0.005 + \frac{1}{p} [0.15 + 0.35 \ (mph/100)^2],$$

in which p is the inflation pressure in lbf/in.^{2} for 27-inch wheels.

Smooth treads on automobile tires reduce rolling resistance by as much as 20 percent, according to Ogorkiewicz (ref. 10); he and Bekker (ref. 2, p. 208) give as an alternative formula for C_R

$$0.0051 + \frac{0.0809 + 0.00012W}{p}$$

$$+ \frac{0.105 + 0.0000154W}{p} \ (mph/100)^2,$$

where W is the weight on the wheel in lbf and p is in lbf/in.^{2}.

Effect of wheel mass on acceleration effort

The wheels of a vehicle move forward with the machine and rider and at the same time rotate around the hubs. The resistance of the wheels to a change in speed is therefore greater, per unit mass, than that offered by the rest of the vehicle. Hence, greater effort is required to accelerate "a pound of weight (mass) in the wheel of a bicycle than a pound in the frame." This fact has been quoted endlessly in cycling literature, both in and out of context.

Bicycle wheels are now of a form such that the major portion of the mass is concentrated in the rim-tire-tube combination. The dimensions of the latter are small compared with the diameter of the wheel, and its center of mass is close to the outside of the wheel, which is traveling at road speed. On this account, it is possible to say with some truth that "the effect of a given mass in the wheels is almost twice that of the same mass in the frame" as far as acceleration power requirements are concerned, because the wheel has to be given both the translational kinetic energy of the whole machine and its own rotational kinetic energy relative to the bicycle.

With modern bicycle construction the wheels form only about 5 percent of the total mass of machine and rider. Also, the effect of any practical variation in reducing this 5 percent is small, whether by reducing the wheels by size or by material content. At best, it is estimated that the wheel mass can be reduced to 3.5 percent of the total. The reduction effect is, therefore, 1.5 percent. Even if this can be multiplied by 2 (because the mass revolves), the resultant 3 percent effect on acceleration is very small and would not be easy to detect.

More accurate estimations based upon calculations or measurements of the actual moments of inertia of 16- and 27-inch wheels show that the difference in acceleration power is rather less than 1.7 percent.

Although lighter wheels accelerate slightly more quickly for a given power, and have a lower air drag, they also have a larger rolling resistance on smooth roads because of the larger losses at the point of contact (figure 5.2). The decision on whether or not to use small wheels must depend on the duty anticipated for the bicycle, as well as on cost and fashion.

Rough roads and springing

Rough roads affect bicyclists in several ways. The vibration may be uncomfortable and may require the bicycle to be heavier than if it were designed for smooth roads, and there will be an energy loss.

The energy loss depends on the "scale" of the roughness, the speed, and the design of the bicycle. If the scale is very large, so that the bicyclist has to ascend and descend large hills,

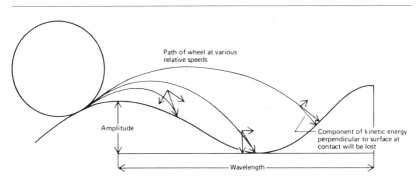

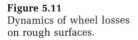

Figure 5.11
Dynamics of wheel losses on rough surfaces.

overall energy losses are small (and due principally to the increased air-resistance losses at the high downhill speeds). There are in this case virtually no momentum losses.

Now imagine a very small scale of roughness, with a supposedly rigid machine traveling over the surface. Each little roughness could give the machine an upward component of velocity sufficient for the wheel(s) to leave the surface (figure 5.11). The kinetic energy of this upward motion has to be taken from the forward motion, just as if the rider were going up a hill. But when the wheel and machine descend, under the influence of gravity as before, the wheel contacts the surface at an angle whose magnitude depends upon the speed and the scale of the roughness. All the kinetic energy perpendicular to the surface at the point of contact can be considered lost. Herein lies part of the reason for rough-road losses.

Pneumatic tires greatly lower the losses for small-scale roughness, because only the kinetic energy of part of the tread is affected, and the spring force of the internal pressure ensures that in general the tire does not come out of contact with the surface. The principal losses are due to the flexing (hysteresis) of the tires and tubes.

At a larger scale of roughness, perhaps with a typical wavelength of 6–60 inches (0.152–1.52 m) and a height amplitude of 1–6 inches (25–152 mm), bicycle tires are too small to insulate

the machine and rider from the vertical velocities induced, and the situation more nearly approaches the analogy of the rigid-machine case
discussed above. For this scale of roughness,
typical of potholes and ruts, some form of
sprung wheel or sprung frame can greatly reduce the kinetic-energy or momentum losses by
reducing the unsprung mass and ensuring that
the wheel more nearly maintains contact with
the surface.

Another way of expressing this conclusion is
that, if energy losses are to be small, the "natural" frequency of the unsprung mass should be
high compared with the forced vibrational frequency imposed by the surface. The natural frequency f_N of a mass m connected to a spring
having a spring constant λ (λ gives the units of
force applied per unit deflection) is

$$f_N = \frac{1}{2\pi} \sqrt{\frac{\lambda g_c}{m}} \text{ vibrations per unit time.}$$

The forced frequency from the road surface, f_f,
is equal to v/S, where v is the velocity of the
bicycle and S is the wavelength of the roughness. Therefore, the ratio

$$\frac{f_N}{f_f} = \frac{S}{2\pi v} \sqrt{\frac{\lambda g_c}{m}}$$

should be kept high by reducing the unsprung
mass m for the worst combination of S and v
thought likely to be encountered. (The designer
has little choice for the spring constant λ, because he must assume a mass of rider and machine of up to perhaps 275 lb (124.7 kg) (having
a weight at sea level of 1,223 N), with a maximum deflection (if a light rider is to be able to
reach the ground with his foot) of perhaps 3
inches (76 mm).

Throughout this book the motion of the bicycle
under consideration has been assumed to be
taking place upon relatively smooth surfaces. In
such circumstances it seems reasonable to as-

sume that energy losses due to vibration are small. Roads are certainly becoming smoother. As a consequence, the task for bicycle designers has been made easier than it was in the earlier days when even in the industrialized societies most of the roads were too rutted for easy riding. In the United Kingdom, where much sport cycling is in the form of time trials, the modern road-racing bicycle is approaching the track bicycle in detail design, as with, for example, small-cross-section lightweight tires. Present-day utility machines are little different in specifications from road racers of the 1920s.

In contrast, the pre-1890s bicycle designer was forced to take serious account of the road surface. An early writer was of the opinion that if the front wheel of a rear-drive safety (fitted with solid rubber tires) was forced to surmount a 4-inch-high obstacle, a loss of half of the forward momentum was experienced.[22] This is an extreme case, but is indicative of the large energy losses likely when riding on very rough roads. It was also known that solid rubber tires were less easy-running as the speed increased, even on relatively smooth roads; the vibration loss is almost directly proportional to speed, even at low speeds. According to Sharp, the French engineer Bourlet thought that one-sixth of the rider's effort was lost to vibratory effects on a solid-rubber-tired bicycle (ref. 18, p. 252).

As can be expected with the above state of affairs, inventors busied themselves with so-called antivibratory devices of all imaginable types.

The application of anti-vibration mechanisms to bicycle frames was difficult. Several designers seemed to have a clear grasp of the essential problems to be solved: The rider must not have to cope with differing distances between saddle and pedals, and forward momentum must be preserved. The general outcome was, however, far from the optimum, and according to Scott (ref. 22)

> . . . the difficulty experienced by inventors on the line of anti-vibrators appears to be, that while acquiring the desired elasticity in the proper direction, an elasticity in other directions has followed, making the machine feel unsteady and capricious, especially in the steering. This undoubtedly valid difficulty in the way is worthy of careful consideration before accepting an anti-vibrator: in fact the very desired end can be easily missed in an imperfect device, as it might, while holding momentum in one direction lose it in another.

In spite of difficulties, inventors persevered and there was some market for machines fitted with a large antivibrator (as distinct from sprung forks or saddles) in the form of a sprung frame. Examples are shown in figures 5.12 and 5.13. The type of frame most praised was the "Whippet." All these machines suffered from wear at the joints, to varying degrees, and what might have been an acceptable machine when new was not so when the joints became loose. The steering of the "Whippet" pattern is seriously affected by wear, as can be surmised by even a casual inspection of the design. To ride a sprung frame that is loose in its essential joints is enlightening and awe-inspiring.

The major deliverance was the invention of the pneumatic tire in 1888. This placed the antivibratory device just where inventors had always wanted it—at the road surface—thus doing away with a chain of actuating connections to the root of energy absorption. At first the pneumatic tire was almost impractical because of its susceptibility to road litter, but development was rapid, and by 1892 most new bicycles were sold with pneumatic tires (although they cost much more than solid or hollow rubber). An early text on bicycles warned that pneumatics were prone to roll on cornering and thus could cause fear to less intrepid riders.[23] Maybe this fact and the tires' fragility delayed their univer-

Figure 5.12
Whippet spring-frame
bicycle. From reference
18.

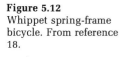

Figure 5.13
Humber spring-frame
bicycle. From reference
18.

sal acceptance among nonracing riders by a year
or two. For road use, it appears that the early
pneumatic tires were run at inflation pressures
of 20–30 lbf/in.^{2} (138–207 kPa), which is far too
low for confident cornering. Probably these low
pressures were thought to avoid straining and
splitting the covers, although they must have
made punctures more likely.

The designer of the "Whippet" frame is
thought to have been convinced that there was
no future in large-scale adaptation of springs to
bicycles after the introduction of the pneumatic
tire. These sentiments were not shared by other
innovators, however; even the large Humber
concern thought that there was a demand for a
sprung-frame bicycle, though with pneumatic

tires (figure 5.12). Over the following decades, this example was followed by others who incorporated pneumatic and other unusual springing. Some of these designs may have been inspired by the light motorcycles that appeared in the twentieth century. One developer was Air Springs Ltd., who marketed a telescoping pneumatic saddle pillar designed by the renowned "Professor" Archibald Sharp.[24] No doubt such sprung machines could have been useful on very rough roads, but improvements in the average road conditions for bicycling were decreasing the need for major springing devices.

In the less-developed parts of the world, where bicycles are ridden in great numbers, the roads are still rough. The most common bicycle is one fitted with large-diameter tires, about $28 \times 1\frac{1}{2}$ inches. This gives tolerable comfort without the use of a sprung frame.

The appearance of a successful modern sprung bicycle (figure 5.14) would seem to contradict the above arguments. However, the logic of the designer, Alex Moulton, was as follows (see ref. 25): For a bicycle to be truly useful to the "utility" cyclist, there has to be better provision for the carrying of luggage than can be fitted to standard machines. If the wheels were made much smaller, room would be created over them. Small wheels would lead to unacceptable vibration and energy losses, especially with "dead" loads such as luggage over them, so springing would be required. Small wheels also make the bicycle a little shorter, so that it can be fitted into the trunk of a standard European automobile. The rear-wheel spring of the Moulton bicycle uses rubber in compression and shear; the front wheel has a rubber-damped coil spring. The resulting bicycle is very effective over both smooth roads and roads too rough for regular bicycles to tackle at any but very low speeds.

A very successful though noncommercial design for a sprung bicycle is Dan Henry's modifi-

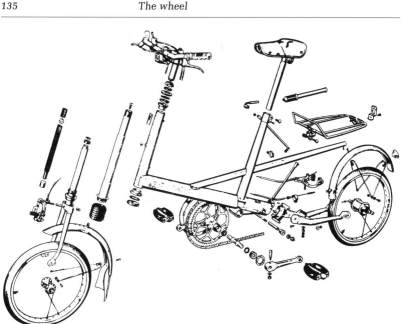

Figure 5.14
Moulton bicycle, with
sprung front and rear
wheels. Courtesy of
Raleigh Industries, Inc.

cation of a lightweight sport machine (figure
5.15). Each wheel is mounted in a swinging fork
on stiff bearings, which gives lateral rigidity
with long up-and-down travel. The springs are
quickly adjustable to the rider's weight. The
wheelbase is longer than that of the standard
machine because of the placement of the rear
wheel, but the steering geometry is unaltered
(with the front wheel in its mean position) because the original front forks are set back to
compensate for the forward set of swinging
forks. Henry has ridden over 100,000 miles on
this machine, which weighs 28 lb (12.7 kg). In
contrast with reports on other sprung bicycles,
he finds that he climbs hills more quickly than
with an unsprung machine and he reports that
the tires last longer. Lightweight sew-up tires
can be used on roads that would require
clincher (wired-on) tires on an unsprung
bicycle.

Figure 5.15
Dan Henry's spring-frame
bicycle. (top) Detail of
the front suspension;
(bottom) normal position.
Courtesy of Dan Henry.

Shape and resilience of spoked tension wheels

Arguments about how a spoked tension wheel supports the load on the axle, about the shape the wheel takes under load, and about the effect of different patterns of spoking on the wheel resilience were reopened when Forester measured rim deflection under load and found it to be negligible compared with the tire deflection.[26] (Forester found Sharp's treatment in ref. 24 excellent as far as it went.) Computer stress calculations and sophisticated measurements on actual wheels were made by various people (see the ensuing correspondence in the columns of *American Wheelmen*). At the time of writing, this discussion has not been fully published and a consensus has not been reached. The following, therefore, represents only the views of D.G.W.

Under load, a spoked wheel takes up not an oval shape, as is often stated, but an approximately circular shape with a flattened portion in the vicinity of road contact. It is thus analogous to a pneumatic tire. Let us picture an inflated tire, mounted on a rim, and consider the tensions in the tire cords as analogs to the tensions in the spokes. The wheel is not held off the road, as is popularly supposed, by air pressure, which is uniform around the wheel rim and therefore contributes zero net force. Rather, the air pressure puts the tire cords into tension, and these cords pull on the wheel rim to provide support. The tire cords may have a bias (an angle to the radial direction), jut as tangent spokes have a bias, but for present purposes let us think of the cords and the spokes as being radial. Both can withstand considerable tension, but not compression.

When the tire is pushed in at the point of contact with the ground, the air pressure is increased, and this increases the tire-cord tension all around the rest of the wheel. However, in the region of the flat spot the tire-wall curvature is greater, and we know from formulas for pressure vessels (such as those in reference 27) that the tension will decrease. Also, the cord acts at an increased angle to the rim because of the bulge that is formed, further reducing the effective radial force in this region.

In the spoked wheel, the increased compressive stress in the rim (analogous to the increased air pressure in the tire) increases the tension of all the spokes except those in the (slightly) flattened region, where the spoke tension naturally decreases. The load on the axle is taken, then, by the combined effect of the increased spoke tension at the top of the wheel and the decreased tension in the region of contact. All other spokes have approximately equal tension (if we assume a symmetrical rim) and balance each other.

References

1. J. L. Koffman, Tractive resistance of rolling stock, *Railway Gazette* (London) (6 November 1964): 889–902.

2. M. G. Bekker, *Theory of Land Locomotion* (Ann Arbor: University of Michigan Press, 1962), pp. 209, 214.

3. E. Barger et al., *Tractors and their Power Units* (New York: Wiley, 1952).

4. J. C. Trautwine, *The Civil Engineer's Reference Book,* 21st edition (Ithaca, N.Y.: Trautwine, 1937).

5. J. Hannah and M. J. Hillier, *Applied Mechanics* (London: Pitman, 1962), p. 36.

6. O. Reynolds, Rolling friction, *Philosophical Transactions* 166 (1876): 155–156.

7. I. Evans, The rolling resistance of a wheel with a solid rubber tire, *British Journal of Applied Physics* 5 (1954): 187–188.

8. V. Steeds, *Mechanics of Road Vehicles* (London: Illiffe, 1960).

9. P. Irving, *Motorcycle Engineering* (London: Temple, 1964), p. 10.

10. R. M. Ogorkiewicz, Rolling resistance, *Automobile Engineer* 49 (1959): 177–179.

11. E. G. McKibben and J. B. Davidson, Transport wheels for agricultural machines, part II, *Agricultural Engineering* 20 (1939), no. 12: 469–475.

12. E. G. McKibben and J. B. Davidson, Transport wheels for agricultural machines, part III, *Agricultural Engineering* 21 (1940), no. 1: 25–26.

13. E. G.McKibben and J. B. Davidson, Transport wheels for agricultural machines, part IV, *Agricultural Engineering* 21 (1940), no. 2: 57–58.

14. R. E. Sabey and G. N. Lupton, Photographs of real contact area of tires during motion, report LR 65, Road Research Laboratory, Ministry of Transport (U.K.), 1967.

15. G. M. Carr and M. J. Ross, The MIRA Single-Wheel Rolling-Resistance Trailers, Motor Industries Research Association, Nuneaton, England, 1966.

16. *Kempe's Engineer's Year Book*, vol. 11 (London: Morgan, 1962).

17. P. D. Patterson, Pressure problems with cycle tires, *Cycling* (28 April 1955): 428–429.

18. A. Sharp, *Bicycles and Tricycles* (London: Longmans, Green, 1896/Cambridge, Mass.: MIT Press, 1977).

19. F. R. Whitt, Power for electric cars, *Engineering* (London) 204 (2 October 1967): 613.

20. G. L. Fowler, Fighting rolling resistance in tires, *Machine Design* (11 January 1973): 30–34.

21. C. R. Kyle and W. E. Edelman, Man-powered-vehicle design criteria, Third International Conference on Vehicle-System Dynamics, Blacksburg, Va., 1974.

22. R. P. Scott, *Cycling Art, Energy and Locomotion* (Philadelphia: Lippincott, 1889).

23. Viscount Bury and G. Lacy Hillier, *Cycling*, third revised edition, *Badminton Library of Sports and Pastimes* (London: Longmans, Green, 1981).

24. G. L. Fowler, Air springs, *Cycling* (23 November 1910): 504.

25. A. Moulton, The Moulton Bicycle, Friday Evening Discourse, Royal Institution, London, 23 February 1973.

26. J. Forester, Held up by downward pull, *American Wheelmen* (August 1980): 13, 14.

27. R. J. Roark and W. C. Young, *Formulas for Stress and Strain*, fifth edition (New York: McGraw-Hill, 1976).

Recommended reading   G. R. Shearer, The rolling wheel—the development of the pneumatic tyre, *Institute of Mechanical Engineers Proceedings* (U.K.) 191 (November 1977).

6   Mechanical friction

Power losses in chain transmission

The retarding effects of wind, road, and gradient have been discussed above. Another resistance to the progress of a bicycle (a far less important one) is that due to friction power absorption by the chain transmission and the bearings. No estimates of these pedal-power requirements have been included in calculations made up to this point.

The loss of power in an automobile transmission can be as high as 15 percent (ref. 1, p. 316). Most of this loss occurs in the transmission itself and the differential, both of which are oil-immersed sets of gears operating at relatively high speeds. The efficiency of a good clean chain can be as high as 98.5 percent (ref. 1, p. 128; ref. 2). The loss of only 1.5 percent is very small in comparison with the power consumption of the wind and road resistances opposing a bicycle's motion. For example (see table 5.7), at 12.5 mph (5.59 m/sec), when 0.074 hp (55 W) is needed to overcome both wind and road resistances, only 0.001 hp (0.75 W) is absorbed by the transmission. The tire rolling resistance, 0.0295 hp (22 W), cannot be estimated to this degree of accuracy (0.001/0.0295 hp, or 3 percent), let alone the power absorbed by the wind. It appears reasonable, therefore, to refrain from including machinery losses in graphs of power usage for bicycle riding.

The predominance of front-wheel drive in the early days of bicycles is understandable in view of the simple, lightweight, and 100-percent-efficient transmission of power from the pedals. However, the disadvantages of this drive system are serious at speeds higher than a few miles per hour. The wheel must be made as large as possible to give "high gears," and this—along with the limited steering arc of the wheel and

the need for applying torque to the handlebars
to resist the pedaling torque—made the machine
difficult for the less acrobatic to master. The addition of gear trains or the use of levers complicated the inherently simple type of drive and
made it less attractive on this account.

Some details of the evolution of modern chain
design are given in references 1–7. Chain-driven
bicycles were first used on very rough roads.
This environment, along with the Victorian passion for cast iron, appears to have influenced
chain and chainwheel design. "Open link"
chains with thick and wide teeth on the cogs
(partly because of the low strength of cast iron)
were common. It was said that road grit
dropped more easily through the big spaces between the links. The small number of teeth led
to rough running because of the variation in the
speed of the chain (as much as 6 percent) as it
passed over a constant-speed cog.

Later, gear cases (oil-bath chain and cog enclosures) became common, even for racing machines, until the roads improved. Smaller-pitch
chains then came into use, with improved running characteristics; there was typically about 1
percent variation in output speed with constant-
speed drive. The precise shape of teeth has been
the subject of much experiment; a modern opinion on the optimum design, credited to Renold,
is given in references 2 and 3. This design uses
an angle of 60° between the flat faces of two
teeth, with circular arcs to the root and to the
tips. The exact nature of these curves is even
now the subject of much discussion from the
viewpoint of world standardization, and technical committees have not agreed on the precise
form.[5,6]

Illustrations of gearwheel and chainwheel
teeth on advertising posters (even those intended for engineering exhibitions) often show
evidence of artistic license in the ugly, inopera-
ble tooth shapes. Not all such errors escape criticism; the celebrated poster artist Toulouse-

Lautrec once lost a commission because his rendition of a chain set was outrageously incorrect in the manufacturer's eyes.

Bearings

Power losses due to bearing friction

For a long time (at least since 1898), the retarding effect of the friction of the standard ball bearings in bicycles has been considered very small. Sharp quoted Rankine as stating that the friction forces amount to one thousandth of the weight of the rider (ref. 4, p. 251). For a 150-lbf (68-kg) rider on a 30-lbf (13.6-kg) bicycle, this means a resistance of 0.15 lbf (0.667 N)—equivalent to an incremental rolling-resistance coefficient of 0.001. The tire rolling-resistance coefficient alone is not known to better than 0.001, and the wind resistance is much less certain; hence, it appears reasonable to disregard the bearing resistance. However, it is interesting to compare this incremental rolling-resistance coefficient of 0.001 with later relevant information, such as that for railway rolling stock given in references 1 and 8. The wheel-plus-bearing rolling resistance is given there as a few lbf per ton, which might be interpreted as a coefficient of 0.002. Of this, the bearing friction alone is probably under 0.001. The power loss in the complete transmission of an ergometer was given in reference 9 as being as low as 5 percent. According to reference 1, chain power losses probably average 2.5 percent. The bearing losses can thus be taken as $5 - 2.5 = 2.5$ percent. At power inputs to a bicycle of 0.12 hp (89 W) and 0.37 hp (276 W), which represent speeds on the level of 12 and 20 mph (5.36 and 8.94 m/sec) for a touring-type machine with an upright rider, the total opposing forces can be calculated as 3.75 and 7 lbf (16.68 and 31.14 N), respectively. The frictional opposing force of 0.15 lbf (0.67 N) given by Rankine is thus expressible as $0.15/3.75 \times 100$ (about 4 percent) at 12 mph and $0.15/7 \times 100$ (about 2 percent) at

Table 6.1 Coefficients of friction of various bearings.

	Coefficient of friction
Annular ball bearing	0,00175[a] 0.0005–0.001[b] 0.001–0.0015[c] 0.0015[d]
Small needle-type roller bearing	0.005[c]
Plain gunmetal bearing, well lubricated	0.002–0.015[e]
Plain metal machine-tool bearing slow-running fast-running	 0.1[f] 0.02[f]
Nylon 66 dry, on nylon dry, on metal lubricated	 0.2[g] 0.07[g] 0.14[g]
PTFE	0.1–0.14[h]
Bicycle-type ball bearing	0.01[i,j]

a. 1-inch balls; data from R. P. Scott, *Cycling Art, Energy and Locomotion* (Philadelphia: Lippincott, 1889), p. 175.

b. 1-inch balls; data from ref. 10 (supplement II).

c. Source: ref. 1 (vol. 1, p. 1242).

d. Source: ref. 2 (p. 48).

e. Source: ref. 2 (p. 49).

f. Source: *Mechanical World Year Book* (Manchester, England: Emmott, 1938), p. 442.

g. Source: *British Plastics* (February 1966): 80.

h. Source: leaflet from Glacier Metal Co., Ltd., Alperton, Wembley, England.

i. Source: A. Sharp, *CTC Gazette* (October 1898): 493, efficiency data of Mr. Carpenter.

j. F.R.W.'s experimental work with $\frac{1}{4}$–$\frac{1}{8}$-inch balls in angular contacts of 30° and 60°, with a $\frac{3}{4}$-inch-diameter running circle, gives an average of 0.01 for radial or thrust loads.

Note: The bicycle-type bearings are assumed to be in very good condition and carefully adjusted. Otherwise, the friction can be several times greater. Poor manufacture can also give such variations.

20 mph; the average of the two cases is about 3 percent.

Information given in table 6.1 shows that the coefficient of rolling friction attributable to 1-inch-diameter (25.4 mm) balls in a bearing is about 0.0015, and reference 10 shows that the rolling friction varies inversely as the ball diameter. We can assume that the typical average contact angle of the bearing is 45°, which increases the effective load on the races by $\sqrt{2}$, or 1.41. Hence, the average bicycle ball bearing, with $\frac{1}{8}$–$\frac{1}{4}$-inch (3.175–6.35-mm) balls, should have a coefficient of $0.0015 \times \sqrt{2} \times \frac{3}{16} = 0.011$. Experimental data are given in figure 6.1.

The effective resistance at the road of the load-bearing wheels is very small because of the large ratio of wheel diameter to bearing diameter. When the losses of the pedals and the

Figure 6.1
Test results for bicycle ball bearings. Line A: rolling-friction coefficient 0.0015. Line B: rolling-friction coefficient 0.001. (Both these lines are for a 1-inch-diameter face with 45° angular thrust.) Line C: Range given by Bourlet (reference 7, p. 15) for 6-mm-diameter balls.

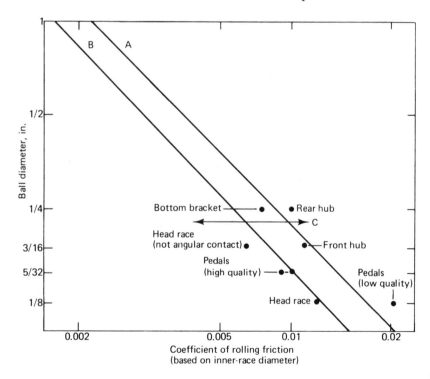

brackets are added to those of the wheels, the overall loss of the pedaler's energy due to friction in the bearings is found to be less than 1 percent. Chain losses would add 1.5–5 percent.

Advantages of ball bearings
The first ball bearings were a far cry from the highly reliable product of today, but they were soon adopted by bicycle makers. Only a few years elapsed before the plain bearings of the "boneshaker" period were abandoned in favor of the more complicated ball bearings. More of the ball-bearing patents during this early period of bearing history were directed toward bicycles than toward any other purpose.

The common cup-and-cone bearing—which is inexpensive and can tolerate some misalignment (a very desirable characteristic for inaccurately made and somewhat flexible bicycles)—appeared as early as the 1880s (see figures 6.2 and 6.3).

Ball bearings require a low starting torque, whereas plain bearings generally require a high starting torque (see reference 11 and figure 6.4). This phenomenon is well appreciated in railway practice. It is now accepted that the use of roller bearings in trains reduces the starting power needed severalfold in comparison with well-lubricated plain bearings, although the running power needed is similar (ref. 1). Plain bearings are sensitive to load and rotation rate because of the changing characteristics of the lubricant film separating the shaft and the bearing. Figure 6.5 shows that under optimum conditions very good performance can be obtained from plain bearings, but the range of coefficients of friction is large for variable conditions of bearing load and speed. (This bearing would have some degree of "hydrostatic" lubrication, which requires considerable power from an external oil pump and therefore would not be applicable to a bicycle.)

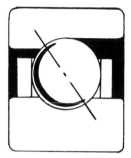

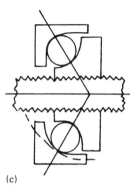

(a)   (b)   (c)

Figure 6.2
Types of ball bearings.
(a) Annular or radial.
(b) 1893 "Magneto" (the
Raleigh version had a
threaded inner race).
(c) Cup-and-cone (from
the diagram it can be
seen that the bearing is
self-aligning and can
accommodate a bent
spindle).

Figure 6.3
High-quality hub with
sealed ball bearings.
Courtesy of Phil Wood.

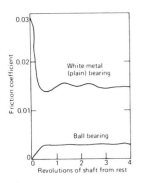

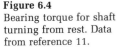

Figure 6.4
Bearing torque for shaft
turning from rest. Data
from reference 11.

If a plain bearing is not kept well lubricated, the friction can increase manyfold. A ball bearing, on the other hand, may deteriorate more quickly without lubrication (particularly if water and grit are allowed to enter), but the friction will not increase greatly. Therefore, it is desirable to continue the use of ball bearings in bicycles.

Some think that plain bearings made from nonmetallic materials could now be used. It has been found that such bearings function in wet conditions and without oil—desirable features for bicycle bearings. Nylon is one nonmetallic bearing material; another is polytetrafluoroethylene (PTFE), a highly corrosion-resistant synthetic polymer. Special PTFE bearings incorporating metal mixtures in order to resist the seizure experienced with pure PTFE have been tested (ref. 1). It appears that the coefficients of friction are 0.10–0.16 for suitable loading and design. Table 6.1 gives information about other bearing materials. The minimum value is still high: 0.04, or several times that associated with ball bearings.

It is probable that with nonmetallic bearings the power needed to propel a bicycle and rider at 10 mph (4.47 m/sec) would be about $1\frac{1}{3}$ times

Figure 6.5
Friction coefficient of a
plain bearing, from G. F.
Charnock, *The
Mechanical Transmission
of Power* (London:
Crosby Lockwood, 1932),
p. 30. This very efficient
bearing, which had a
steel shaft in a rigid ring-
oiling 3-inch-diameter
pillow block with
gunmetal steps and was
lubricated with
"Gargoyle Vaculine C,"
was probably about four
times as easy-running as
an average plain bearing.

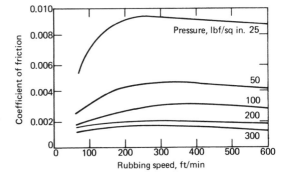

that needed with ball bearings, on the level and in still air. This estimation assumes that the bearing-friction effect would be ten times that associated with ball bearings, which gives an effective rolling-resistance coefficient of 0.01. It appears, therefore, that nonmetallic bearings would be suitable only for, say, machines intended for children or certain invalids whose speed it might be desirable to restrict for safety purposes.

Life of bearings

Although the life of a plain bearing in a turbine (for instance) is virtually infinite, because high-pressure lubrication and high-velocity relative motion combine to prevent metal-to-metal contact, such conditions could not be approached in a bicycle bearing. Short life and high friction must be expected. On the other hand, ball bearings always have a limited life, but the life can be adequate. The time between overhauls of many aircraft turbine engines is well over 20,000 hours, and the bearings are not usually changed. Cup-and-cone ball bearings on bicycles are made of inexpensive steels, inaccurately constructed, and little protected from grit, and can be expected to need replacement after 1,000 hours. However, some specialty manufacturers are supplying wheel hubs incorporating standard automobile-type ball-bearing assemblies to achieve lower friction, longer life, and less maintenance.

The Sturmey-Archer type of hub gear is an exception to the suggestions that cup-and-cone bearings and plain bearings have short lives in bicycle use. Effective labyrinth dirt seals are used; the balls are enclosed in cages that eliminate ball-to-ball rubbing; and bearings are accurately aligned. Early Sturmey-Archer gears (in 1909) incorporated ball bearings in the mounting of the pinion gears. This was claimed to eliminate 60 percent of the friction. However, the bearing loads on these pinion mountings are extremely low, and plain bearings (hardened-steel pins) were substituted without comment later in 1909 and appear to give an acceptable life. The actual (rather than the claimed) effect on the gear efficiency of this substitution is not known.

Friction in the rider's limb joints

Human limb joints are "plain bearings," lubricated by a natural fluid that exudes from the bearing surface. Recent investigations assign a average coefficient of friction to this very special type of bearing.[12] Bourlet quotes M. Perrache as stating that the joint friction work done by a pedaler during one revolution of the crank set as the equivalent of 3.9 joules.[13] A cyclist thrusting hard on the pedals will do about 200 joules' work per crank revolution. Hence, the pedaler loses about 2 percent of the pedaling effort in overcoming joint friction.

References

1. *Kempe's Engineer's Year Book,* vol. 11 (London: Morgan, 1962).

2. G. F. Charnock, *The Mechanical Transmission of Power* (London: Crosby Lockwood, 1953).

3. R. F. Kay, *The Theory of Machines* (London: Arnold, 1952), p. 278.

4. A. Sharp, *Bicycles and Tricycles* (London: Longmans, Green, 1896 / Cambridge, Mass.: MIT Press, 1977).

5. British Standards Institution Association publication B.S228, 1954.

6. American Standards Institution Association publication 1329.

7. C. Bourlet, *La bicyclette, sa construction et sa forme* (Paris: Gauthier-Villars, 1889), pp. 85–97.

8. J. L. Koffman, Tractive resistance of rolling stock, *Railway Gazette* (London) (1964): 899–902.

9. W. von Döbeln, A simple bicycle ergometer, *Journal of Applied Physiology* 7 (1954): 222–229.

10. L. Levinson, in *Fundamentals of Engineering Mechanics,* ed. J. Klein (Moscow: Foreign Languages Publishing House, 1968).

11. *A Dictionary of Applied Physics*, ed. R. Glaze-brook (London: Macmillan, 1922).

12. S. A. V. Swanson and M. A. E. Freeman, Mechanism of human joints, *Science* (February 1969): 73–78.

13. C. Bourlet, Le nouveau traité des bicycles et bicyclettes le travail, in *Encyclopédie scientifique des aide mémoire*, ed. M. Léaute, second edition (Paris: Gauthier-Villa s, 1898), p. 146.

Recommended reading

F. P. Bowden and D. Tabor, *Friction and Lubrication* (London: Methuen, 1956).

C. F. Caunter, *Cycles—A Historical Review* (London: Science Museum, 1972). (This contains a good review of various types of gearing.)

D. Swann, *The Life and Times of Charley Barden* (Leicester: Wunlap, 1965), p. 58.

The relationship between power and speed

Having reviewed the power-output capabilities of humans and the various power losses associated with bicycles and similar vehicles, we can now combine these characteristics to arrive at the power requirements for traveling at various speeds on different types of bicycles. We can also place bicycling along the entire range of muscle-powered movement, and compare it with other modes of wheeled transportation (such as roller skating) and with walking. And we can give a scientific answer to a question that repeatedly raises itself to the touring cyclist in hilly country: When is it better to dismount and walk up a hill than to continue straining on the pedals?

It is easy to show that the bicycle is very efficient. However, to claim that it is even more efficient than the dolphin—a frequently heard extravagance—is to make an unscientific statement. The resistance to motion, and therefore the overall energy-efficiency, is a strong function of speed for all modes. The way in which the resistances vary with speed is peculiar to each vehicle, animal, or mode. Therefore, comparisons are valid only if they are made at the same speed. The bicycle still comes out well.

Figure 7.1 shows the world-record speeds for different durations for the principal forms of human-powered propulsion. Presumably the contestants were putting out about the same power in each mode for the same durations. The standard lightweight track bicycle is 4–8 mph faster than the best speed skater. The astonishing jump in record speeds from standard racing bicycles to machines using streamlined fairings in the IHPVA races adds another potential advantage to bicycling.

Figure 7.1
World-record speeds by human power in various modes. Point at top and dashed curve for unlimited-class vehicles in IHPVA trials.

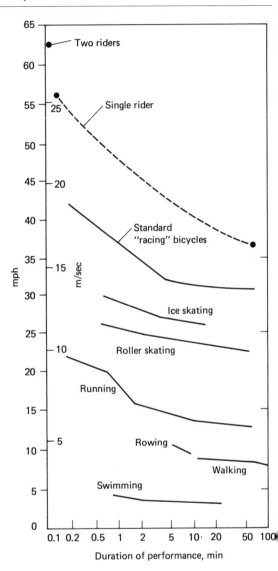

Table 7.1 Principal bicycling speed and time-trial records.

World's track records

Professional unpaced standing start:
1 km, Milan, 1952, R. H. Harris, 1 min 8.6 sec
1 h, Mexico, 1972, E. Merckx, 49.408 km

Amateur unpaced standing start:
1 km, Mexico, 1967, G. Sartori, 1 min 4.6 sec
1 h, Mexico, 1969, T. Radames, 46.95 km

Unofficial and unrestricted:
1 h, motor paced, standing start, Montlhéry, 1928, L. Vanderstuyft,
76 miles, 503 yards
1 km, motor-paced, flying start, Freiburg, 1962, J. Meiffret,
127.25 mph (204.77 km/h)
1 mile, motor-paced flying start, U.S.A., 1973, A. V. Abbott, 138.67 mph

British amateur unpaced road records

Time trials		h	min	sec
Men				
Bicycle:				
25 miles	A. R. Engers, 1978	0	49	24
100 miles	P. W. Griffin, 1978	3	45	28
12 h	E. J. Watson, 1969	281.87 miles		
24 h	R. Cromack, 1969	507.00 miles		
Tricycle:				
25 miles	D. Worsfold, 1978	0	57	38
100 miles	A. J. Pell, 1975	4	25	45
12 h	H. Bayley, 1966	249.65 miles		
24 h	E. Tremaine, 1972	457.89 miles		
Women				
Bicycle:				
10 miles	B. Burton, 1973	0	21	25
100 miles	B. Burton, 1968	3	55	5
12 h	B. Burton, 1967	277.25 miles		
24 h	C. Moody, 1969	427.86 miles		
Tricycle:				
10 miles	L. J. Hanlon, 1977	0	20	49
100 miles	J. Noad, 1975	5	11	08
12 h	J. Blow, 1960	212.82 miles		
24 h	J. Blow, 1969	374.15 miles		
Tandem bicycle				
30 miles	J. Pitchford and C. M. Goodfellow, 1973	1	4	32

Table 7.1 continued

Long distance		days	h	min
1,000 miles				
Bicycle	R. F. Randall, 1960	2	10	40
Tricycle	A. Crimes, 1958	2	21	37
Tandem bicycle	P. M. Swinden & W. J. Withers, 1964	2	18	9
Tandem tricycle	A. Crimes & J. F. Arnold, 1954	2	13	59
Land's End to John o' Groats (872 miles)				
High ordinary	G. P. Mills (P), 1886	5	1	45
Bicycle	R. F. Poole, 1965	1	23	46
Tandem bicycle	P. M. Swinden & W. J. Withers, 1966	2	2	14
Tricycle	D. P. Duffield, 1960	2	10	58
Tandem tricycle	A. Crimes & J. F. Arnold, 1954	2	4	26

Source: Cyclists' Touring Club

Note: Record speeds for vehicles racing under the rules of the International Human-Powered Vehicle Association (IHPVA) are shown in figure 4.9.

World records achieved on standard bicycles are listed in table 7.1. These speeds may be derived reasonably accurately from the maximum power outputs of athletes for various durations (figure 2.10), the air-drag and rolling-friction-drag figure (chapters 4 and 5), and an estimate of the other frictional resistances in the transmission and the wheel bearings (chapter 6). The air-drag and rolling-power requirements of unstreamlined bicycles and tricycles are plotted in figure 7.2.

Figure 7.2
Power requirements for
propulsion of bicycles
and tricycles.

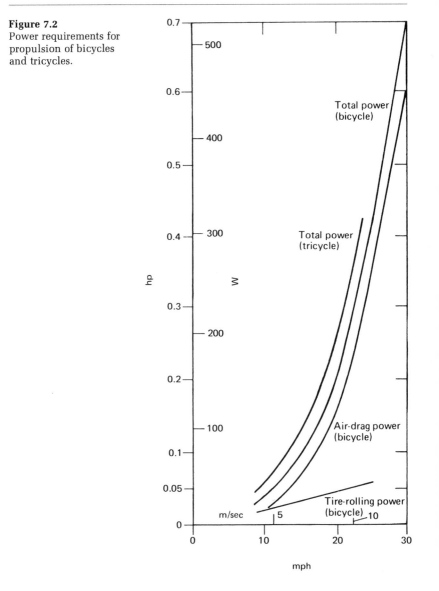

Effect of bicycle type on performance

We have performed calculations for three present types of bicycles and two hypothetical future vehicles. The "slowest" of the present cycles was a three-speed "roadster," on which the rider sits almost upright, presenting almost the maximum area in a rather high-drag shape, flat-on to the wind. Although this type was known a decade or so ago as the "English lightweight," it is light only in comparison with the old heavy-duty single-speed American-style bikes, which have almost disappeared from the catalogs. The second bike we considered was the 10-speed "sports" type, which has dropped handlebars so that the rider (when using the lower handgrips) not only presents a smaller frontal area but also assumes a lower-drag shape. The sports bike is lighter than the roadster and has lighter, smaller-section wheels with higher-pressure tires. We assumed that the rider plus clothing would generally weigh less than would the typical rider plus clothing on a roadster. For the frontal area and drag coefficient shown, we assumed that the rider would typically adopt a crouched position. The third bicycle we took was a true racing model, considerably lighter still than the sports, with very light tubular tires. The rider would adopt an even more "tucked-down" position than for the sports bike and would wear tight clothing. The hypothetical future vehicles considered are called CHPVs (commuting human-powered vehicles) and UHPVs (ultimate human-powered vehicles). The CHPV is hypothesized to be a recumbent bicycle with a streamlined fairing or enclosure, sufficiently high off the ground to give stability and visibility in highway use. The UHPV is a machine to win a future IHPVA 200-m flying-start speed trial. It will be very low to the ground, have a minimum-drag (and therefore long) fairing, and may have two, three, or four wheels.

The power required at the pedals of these cycles at any speed on a level, smooth, hard surface (corresponding to that for which the rolling-resistance coefficient is quoted), in still air, is

$$\dot{W} = \frac{C_V}{\eta_{mech}} \left\{ \Sigma mg \left[C_R + \frac{s}{100} + \frac{a}{g} \left(1 + \frac{m_W}{\Sigma m} \right) \right] \right.$$

$$\left. + 0.5 C_D A \rho (C_V + C_W)^2 \right\},$$

where η_{mech} is the overall mechanical efficiency of the transmission (including wheel-bearing losses), Σm is the total mass of the rider plus clothing plus machine in kg, C_R is the coefficient of rolling resistance, g is the gravitational acceleration (9.806 m/sec² at sea level), C_v is the speed of the bicycle in m/sec, s in the upslope in percent, a/g is the vehicle's acceleration as a proportion of gravitational acceleration, m_W is the effective rotational mass of the wheels and tires at the outside diameter in kg, C_W is the headwind in m/sec, C_D is the aerodynamic drag coefficient, A is the frontal area of rider plus machine in m², and ρ is the air density in kg/m³.

The air density at sea level, at 15°C, is about 1.226 kg/m³. Using a typical overall mechanical efficiency of 0.95, the power equation can be expressed as

$$\dot{W} = C_V \left[K_1 + K_2 (C_V + C_W)^2 \right.$$

$$\left. + 10.32 \Sigma m \left(\frac{s}{100} + 1.01 \frac{a}{g} \right) \right],$$

with the values of the constants K_1 and K_2 for these typical conditions listed in table 7.2, which summarizes the assumptions coming from these considerations. The correction factor for the rotational acceleration of the wheels has been given the typical value of 1.01. The power required for constant-velocity movement on a level road in still air simplifies to

$$\dot{W} = K_1 C_V + K_2 C_v^2.$$

Table 7.2 Specification of vehicle types.

| | Vehicle type[a] | | | | |
| | Present | | | Future | |
	RR	SS	RG	CHPV	UHPV
Frontal area, A	0.5 m²	0.4 m²	0.33 m²	0.5 m²	0.4 m²
Drag coefficient, C_D	1.2	1.0	0.9	0.2	0.12
Bicycle mass, m_b	15 kg	10 kg	6 kg	20 kg	15 kg
Rider + clothing mass, m_c	80 kg	75 kg	75 kg	80 kg	75 kg
Total mass, Σm	95 kg	85 kg	81 kg	100 kg	90 kg
Rolling-resistance coefficient, C_R	0.008	0.004	0.003	0.003	0.002
Constants[b]					
K_1	7.845	3.509	2.508	3.097	1.858
K_2	0.3872	0.2581	0.1916	0.0645	0.03097

a. Vehicle types are designated here and in figures 7.3 and 7.4 by the following abbreviations. RR: roadster bicycle (heavy; upright handlebars). SS: sports bicycle (medium-weight; dropped handlebars). RG: racing bicycle (lightweight; dropped handlebars). CHPV: commuting human-powered vehicle (streamlined fairing; semirecumbent riding position). UHPV: ultimate human-powered vehicle (low, streamlined fairing; recumbent or supine riding position).
b. The constants are used to estimate the propulsion power, in watts, required on a smooth level road having an upslope of s percent, with a headwind of C_W m/sec, to give the vehicle a velocity of C_V m/sec and an acceleration of a m/sec²:
$$W = C_V [K_1 + K_2 (C_V + C_W)^2 + 10.32 \, \Sigma m \, (s/100 + 1.01a/g)].$$

Power curves

The power requirements for moving the three types of bicycles and their riders at various speeds are shown in figures 7.3 and 7.4. The air-drag power is for sea level and 15°C; the rolling-resistance power is for cycling on a smooth, level road. The power to ride up a gentle hill of 2.5-percent slope is also shown in figure 7.3.

It can be seen that for the commuting or shopping bicyclist who travels at around 10 mph (4.5 m/sec), the rolling and air-drag powers are similar (with no wind). Hills and headwinds are very important to these riders. For the higher-speed rider, air drag becomes dominant. One-hour races are usually won at over 25 mph (11.2 m/sec), at which speed the power going to conquer air resistance may be five or more times that being lost in rolling resistance. At these

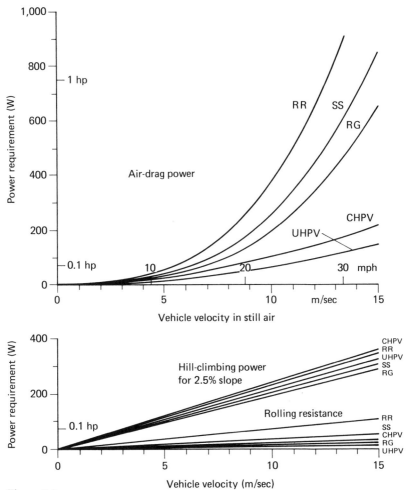

Figure 7.3
Power requirements for bicycling. RR: roadster. SS: sports bicycle. RG: racing bicycle. CHPV: commuter human-powered vehicle. UHPV: ultimate human-powered vehicle. The 0.1-hp output of a typical nonathlete cyclist is indicated.

Figure 7.4
Power requirements for
bicycling. Dashed line
represents output of
typical nonathlete
bicyclist.

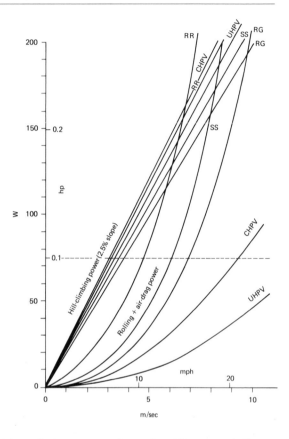

speeds it may even be easier riding up a hill
with the wind behind one than down the hill
with the wind in one's face.

The rolling and air-drag powers have been
added in figure 7.4 for only the lower end of the
speed range. These curves can be accurate only
for still-air conditions. Plotted this way, these
curves show the penalty of riding a roadster
rather than a sports or racing bike. At 10 mph,
about twice as much power is required for the
roadster as for a sports bicycle. Another way of
looking at the difference is to take 0.1 hp (about
75 W) as the output most fit adults feel they
could keep up for hours. At this power level,

the roadster would be going at about 10 mph
and the racer about 15.

For the CHPV, an input of 0.1 hp gives a
smooth-road still-air speed of just over 20 mph
(9 m/sec), which, at this very moderate expenditure of energy, would make commuting much
more attractive for many people.

Let us also use these figures to estimate what
the UHPV's record speed might be in the
IHPVA speed trials with a rider of the capability
of Eddy Merckx, whose one-hour ergometer
power output is shown by figure 2.10 to be considerably above the NASA curve for "first-class
athletes" and above the likely outputs of British
amateur time-trial winners. Let us also hypoth-
esize that new mechanisms with optimum hand
and foot motions will allow all people, including those of Eddy Merckx's caliber, to produce
more power. If the total period of effort required
for the time trial is 60 seconds (53 seconds of
acceleration and 7 seconds in the 200-m
course), we can project a power output during
this period of 830 W. Then, using the values of
the constants in table 7.2 for the UHPV, we can
confidently predict that a single-rider vehicle
will eventually reach 65.4 mph (29.25 m/sec).
Indeed, because none of the figures assumed is
particularly optimistic, this speed may even be
exceeded. During such a record attempt, much
of the rider's output will go to providing the kinetic energy of the vehicle plus rider. At 29.25
m/sec the translational kinetic energy of the hypothesized UHPV (total mass 90 kg) is

$\frac{1}{2}mC^2 = \frac{1}{2} \times 90 \times 29.25^2 = 38{,}500$ joules.

The rotating parts have additional kinetic en-
ergy of rotation equal to $I\omega^2$, where ω is the rotational velocity in radians per second and
where I is the moment of inertia given by $I = mk^2$, where k is the "radius of gyration" (the radius at which the rotational mass can be consid-
ered to be concentrated). Let us suppose that the
mass of each wheel rim and tire and tube is 0.5

kg, concentrated at a radius of 0.325 m, so that the moment of inertia of each wheel is

$I = 0.5 \times 0.325^2 \approx 0.05$ kgm².

The outside diameter of the wheels will be about 0.686 m, so the rotational speed will be

$\omega = 29.25 / 0.343 \approx 85$ rad/sec.

Therefore the rotational kinetic energy will be

$\frac{1}{2}I\omega^2 = \frac{1}{2} \times 0.05 \times 85^2 \approx 180$ J

for each wheel.

If there are three wheels, the total rotational kinetic energy will be about 540 J, adding about 1.5 percent to the translation kinetic energy. Though small, this is not negligible, and it provides some justification for using a two-wheeled configuration, rather than three or four wheels, for the "ultimate" vehicle.

The rotational kinetic energy of the parts of the transmission, such as the chainwheel, cranks and chain, are relatively very small for all human-powered vehicles.

Energy consumption as a function of distance

We can use the specifications of table 7.2 to find the energy consumed in bicycling various distances on level ground.

In the physical sciences energy is measured in joules (1 J/sec = 1 W), but in nutrition kilocalories are used to measure the energy content of food. A kilocalorie is the heat or work energy required to raise the temperature of a kilogram of water one degree Celsius, and is equal to 4,186.8 joules. (Unfortunately, in nutrition it is usually abbreviated to "calorie," which confuses physicists.)

A human being is like a fuel cell, taking in chemical energy in food or fat and putting out work energy. The efficiency is defined as the amount of work energy divided by the chemical or food energy. Now, a human being obviously needs some food energy just to live and keep warm, even if no work or other activity is undertaken. We then define the "net" metabolic efficiency as the ratio of the work output to the *incremental* food-energy intake over that necessary to support life. Values between 20 and 30 percent have been measured for trained athletes (figure 2.18). We have used a reasonable mean number for fit people of 0.2388, or 23.88 percent, because when multiplied by 4,186.8 J/kcal it gives 1,000 in the calculation of figure 7.5. For this value of net efficiency, a consumption of one kilocalorie of food energy produces one kilojoule of work.

In figure 7.5 the true relative air velocity is entered, accounting for head and tail winds, and the results will be the sum of the expenditures of rolling resistance and air drag. The effects of hill climbing can also be added: 0.932 kcal per meter climbed for the roadster, and 0.834 and 0.795 for the sports and racing machines. If rider and bicycle together weigh more or less than the amounts taken as typical in table 7.2, the rolling losses and the hill-climbing expenditures should be multiplied by the ratio of the combined mass to that in the table (1 kg ≈ 2.2 lb). To estimate the effects of different frontal areas, drag coefficients, and rolling-resistance coefficients refer to chapters 4 and 5. Underinflated tires or sandy, slushy, or cobbled roads will give rolling losses much higher than the values calculated here for smooth surfaces and well-inflated tires.

All this information has important implications for the energy crisis. We can see from figure 7.5 that a racing bicyclist at 20 mph could travel more than 1,350 miles per U.S. gallon if there were a liquid food with the energy content of gasoline. (Milk is mostly water, but has enough energy to take a racing bicyclist about 95 miles per gallon, so bicyclists could help to solve America's energy shortage and milk surpluses simultaneously.)

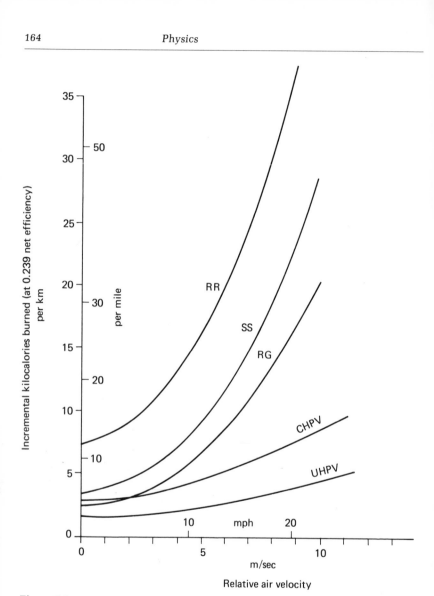

Figure 7.5
Energy consumption in
bicycling over distances,
with equal "incremental"
(net) metabolic
efficiencies assumed.

Confirmation of calculation methods
by recent data

Although some of the typical data and specifications used above (for example, in table 7.2) had their origin in work in France in the 1890s (principally by Bourlet; see reference 1), there has been ample recent confirmation of their value. Nonweiler's data,[2] from the 1950s, are shown in figure 2.5. Moulton[3] and Buckett[4] used sophisticated torque-measuring chainwheels incorporating strain gauges; Kyle et al.[5] used the simple coast-down technique. Their data can be compared with predictions made using the methods of earlier chapters as follows.

Figure 2 of Kyle's report (ref. 5) shows that the power needed to move a crouched rider weighing 73 kg on a Rapido machine weighing 10.7 kg was 97 W at 6.7 m/sec. The power predicted from figure 7.4 and table 7.2 for a rider and machine weighing 81–85 kg is 71–99 W for racing and sports bicycles, respectively. Kyle's power level is predicted exactly by the data on the sports bicycle with the reduced combined weight.

Buckett's thesis (reference 4, figure 6) shows a calculated curve using data as described in chapter 2 and reference 6. At 6.7 m/sec the experimental power required was 127 W, compared with 97 W predicted. Buckett estimated the frontal area, and his machine does not appear to be of a particularly "fast" type. In addition, some additional transmission losses must have occurred. All that can be said is that Buckett produced no evidence to undermine confidence in the data given in figure 2.5 and table 2.2.

Moulton's paper (reference 3, figure 9) gives data on power at the crank for machines with wheels 16, 17, and 27 inches in diameter using an experimental strain-gauge-equipped chainwheel. The rider did not have to read his own instruments (as did Buckett) for speed and torque. This was done via a "fishing line" connection wire. The power at 6.7 m/sec for the average of two machines with 16-inch-diameter wheels is about 194 W, versus 187 W predicted. Moulton gives an estimated frontal area of 0.56 m², which gives more usefulness to the comparison. There is deviation at lower speeds of about 10 fewer watts needed for the machines with 16-inch diameter wheels. However, in view of the unknown transmission losses, it appears that the Moulton figures substantiate the prediction methods given here. Moulton also gives data for a Hetchins (racing) machine with 27-inch wheels. At 6.7 m/sec, 142 W are required. In view of the unknown transmission losses, these results again substantiate the prediction methods. Other data given by Moulton are for a tubular-tired 27-inch-wheel Hetchins and a 17-inch-diameter-wheel machine with nylon-thread tires. The latter showed surprisingly low power requirements compared with all other machines used.

Considering the predictable higher power requirements that were obtained for the 16-inch-wheel machines compared with the single 27-inch-wheel machine as confirming the effect of wheel diameter, a major factor for the good performance of the 17-inch-wheel machine must reside in the nylon-thread tires. DeLong gives a plot of the relative rolling resistances of 27-inch-diameter bicycle tires (figure 8.3 of reference 7) showing that quite surprisingly low rolling resistances, less than half those of standard tires, are obtained from nylon tires.

In contrast with the techniques employed by previous experimenters, Pugh measured the performances of bicyclists pedaling on an ergometer and actually riding on a flat concrete road.[8] The rolling and air friction could be estimated from the oxygen consumptions. Speeds in the road-racing range—27 mph (12.1 m/sec)—were reached. The rolling-resistance and drag coefficients obtained agreed reasonably well with values quoted in the literature by Nonweiler (ref. 2)

Figure 7.6
Relation between oxygen
intake and speed of six
competition cyclists.
From reference 8. Large
points △ calculated from
data on racing cyclist in
reference 6.

Figure 7.7
Relation between net
energy expenditure and
cycling speed for six
competition cyclists.
Inset: plot of energy
expenditure ($\dot{W}$) versus
square of speed ($C_v{}^2$).
Large points △ calculated
from data on racing
cyclist in reference 6.

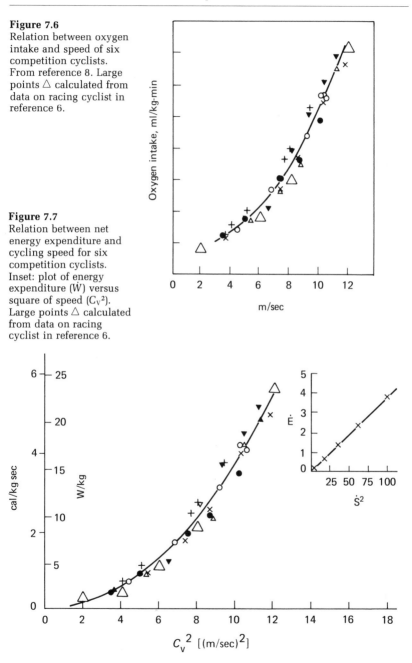

and by Whitt (ref. 6), particularly for concrete surfaces. The oxygen consumptions obtained by Pugh per unit weight of rider for given speeds agreed very closely with those calculable from the table given in reference 6 for racing cyclists.

In figures 7.6 and 7.7, predicted points are superimposed on data from reference 8. Predictions made by the same methods also give good agreement with the work of Hermans-Telvy and Binkhorst[9] and Davies.[10]

Power needed for land locomotion

In order to survive, living species like animals and humans had to develop controllable movement, independent of gravitational and fluid forces which are the usual basis for movement of inanimate objects. The animal world developed "lever systems," which pushed against the ground in various ways from crawling, as do snakes, through bounding, like rabbits, to walking, as practiced by man, which in some ways is like the rolling of a spoked but rimless wheel. With the adoption of the wheel, yet another lever mechanism for movement, came the chance of using a separate inanimate source of power other than that of the muscles of the moving creature. Steam, internal-combustion-engine, and electric vehicles rapidly appeared when lightweight engines of adequate power had been produced.

The bicycle is only one of the many man-developed lever systems for land transport, but it is the sole remaining type that has a limited propulsive power. All other wheeled vehicles have, in general, been fitted with driving units of progressively increased power. In ancient times teams of horses or cattle succeeded single draught animals. The urge for more power and speed seems ever present in human activities.

Animals or wheels

The relative power needed to move a vehicle or animal against ground resistance by various means is shown in figure 7.8. At speeds of a few

Figure 7.8
Power requirements of
human walking and
propulsion of various
animals and vehicles.
Some data from reference
11.

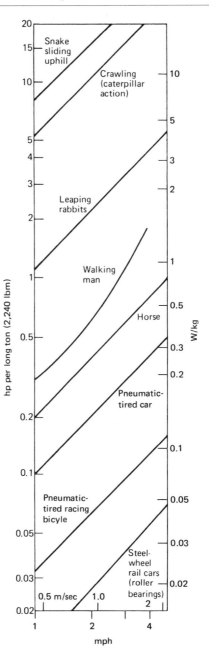

miles per hour these sliding, crawling, leaping, or rolling motions absorb almost all the power exerted by the subject, so that wind resistance can be neglected for purposes of approximate comparison. At higher speeds, the resistance to motion due to air friction assumes a dominant role and obscures the more fundamental difference between wheel motion and other systems of movement based on leverage.

Lever systems are intrinsically efficient, and figure 7.8 (which includes data from reference 11) shows that Nature, in developing walking for man's progression, has given him a system more economical in energy use than that employed by many other animals. Nature has also arranged for her lever systems to be adjusted automatically according to the resistance encountered. The stride of the walker changes, for instance, according to the gradient. In this respect the rider of a bicycle is at a disadvantage, because bicycle gearing that automatically adjusts to give an optimal pedaling rate is not yet perfected. Such a device would have advantages when a high power output over varying conditions was wanted. Modern multigeared bicycles can approximate, if skillfully used, an automatic infinitely variable gear. For low power output, such as is needed for low speeds, the combination of foot pressure and crank revolution rate is not critical.

Bicycles versus other vehicles

The bicycle and rider, in common with most other wheeled vehicles, can move over hard smooth surfaces at speeds at which air resistance is significant—that is, at speeds greater than the 5-mph (2.2-m/sec) upper limit of figure 7.8. The sum total of wind resistance, ground-movement resistance, and machinery friction decides the rate of progress for a given power input to a vehicle. These resistances have been studied carefully over a long period for the commonly used machines, such as those using pneumatic tires on pavement and steel wheels on steel rails.

Graphs showing how the individual resistances contribute to the total for bicycles, railway trains, and automobiles are given in figures 7.2, 7.9, and 7.10. In each case typical examples of vehicles without special streamlining have been chosen in order to bring out reasonable comparisons. The tricycle has been included because it shows the incremental effort needed for propulsion (up to 10 percent above that for the bicycle, as can be deduced from the times achieved in races). Published information concerning the power and performance of mopeds is given in table 7.3.

Figure 7.9
Power requirements for propulsion of 2,240-lbm (1,016-kg) automobile with frontal area of 20 ft² (1.86 m²). Vertical axis represents propulsion power at wheels.

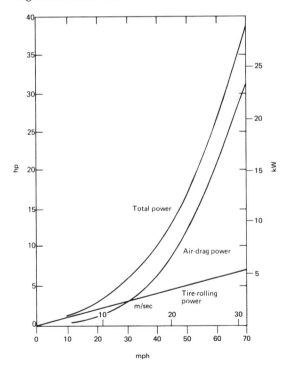

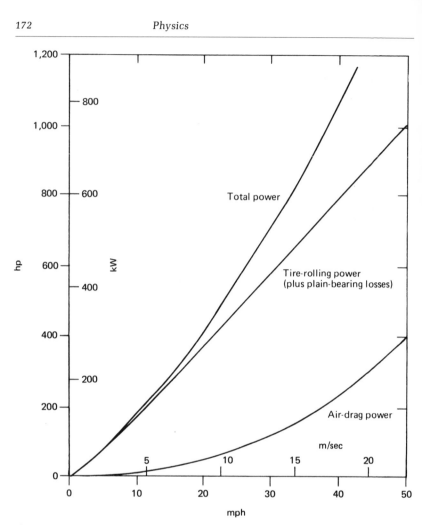

Figure 7.10
Power requirements for
propulsion of freight
train. Data from reference
12, p. 1058.

Table 7.3 Power required to propel mopeds.

Make	Engine data			Moped mass		Rider mass		Level-road max. speed		Wheel diameter	
	hp	kW	rpm	lb	kg	lb	kg	mph	m/sec	in.	m
Powell[a]	1.05	0.78	3,500	—	—	—	—	26	11.6		
Mobylette	1.35	1.0	3,400	75	34	200	91	30	13.4	≈ 26	0.66
Magneet[b]	1.6	1.2	4,700	115	52	200	91	33	14.8		
Raleigh	1.4	1.0	4,500	77	35	182	82.5	26	11.6	16	0.406

Sources:
a. Cycling (9 July 1958): 24.
b. Cycling (27 June 1957): 537.

Our present purpose in comparing these various means of locomotion is to relate the bicycle to other common road vehicles. Some relative power requirements are shown in figures 7.8 and 7.11. Table 7.4 shows that, of all the vehicles, bicycles are impeded the most by wind. A feature of modern automobiles is the relatively high power absorbed by the tires (figure 7.3). In contrast, railway trains are hardly affected by wind resistance below 40 mph (17.9 m/sec) (figure 7.10). With regard to the propulsion power required per unit weight, the bicyclist can be seen to need far less than the walker at low speeds.

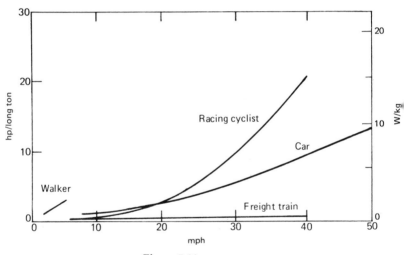

Figure 7.11
Power requirements of human walking and propulsion of racing cyclist, automobile, and freight train over a range of speeds. Data from table 1.2.

Table 7.4 Estimated forces opposing the motion of various vehicles on smooth surfaces in still air (typical cases).

Vehicle and weight	Origin of force	Resisting force, lbf (N)			
		5 mph (2.24 m/sec)	10 mph (4.47 m/sec)	20 mph (8.94 m/sec)	40 mph (17.9 m/sec)
Man walking, 150 lb (68 kg)	Wind	0.2 (0.89)			
	Rolling	13.0 (57.9)			
	Total	13.2 (58.7)			
Cyclist, 170 lb (77 kg) (racing type)	Wind	0.2 (0.89)	0.8 (3.6)	3.2 (14)	12.8 (57)
	Rolling	0.9 (4.0)	0.9 (4.0)	0.9 (4.0)	0.9 (4.0)
	Total	1.1 (4.9)	1.7 (7.6)	4.1 (18)	13.7 (61)
Auto, 2240 lb (1016 kg)	Wind	0.9 (4.0)	3.5 (15.6)	14.0 (62)	56.0 (249)
	Rolling	37.0 (167)	37.0 (165)	37.0 (165)	37.0 (165)
	Total	37.9 (169)	40.5 (180)	51.0 (227)	93.0 (414)
Freight train, 1,500 tons	Wind	35 (156)	140 (620)	560 (2490)	2,250 (10,010)
	Rolling	7,500 (33,370)	7,500 (33,370)	7,500 (33,370)	7,500 (33,370)
	Total	7,535 (33,530)	7,640 (33,990)	8,060 (35,860)	9,750 (43,380)

Human versus animal muscle power

The power available for propelling a bicycle is limited to that of the rider. Let us study how human muscle power compares with that of other living things with similar muscle equipment.

For thousands of years—and even today in the less-developed parts of the world—horses, cattle, dogs, and humans have been harnessed to machines to turn mills, lift water buckets, and do other tasks. When the steam engine was invented, it was necessary to have handy a comparison between its power and that of a familiar source. Experiments showed that a big horse could maintain for long periods a rate of lifting power equal to that of raising 33,000 lb (14,698 kg) one foot (0.3048 m) in one minute. This value came to be universally accepted as the "horsepower." Average horses could in fact work at a greater rate, but only for briefer periods which were not useful. Reference 12 expands on the relationships between total output per day and rate of output.

Other information relating peak power output to duration of effort is given in table 7.5 and figure 2.5. It seems that a man tends to adjust his power output to rather less than 0.1 hp (74.6 W) if he intends to work for other than very short periods and is not engaged in competition. This power level can be shown by experiments and by calculation (figure 7.4) to move a bicyclist and machine on the level at 9–15 mph (4.0–6.7 m/sec), depending on wind resistance, type and weight of bicycle, and condition of road surface. This range of speeds has been associated with average cycling since the standardization of good rear-driven pneumatic-tired bicycles.

Recently the breathing rates of pedaling bicyclists have been measured. Reference 13 describes such experiments with riders moving at 10 mph (4.47 m/sec) and using 0.1 hp (74.6 W). Reference 14 shows that at about this power

Table 7.5 Power outputs of horse and man.

	Period	hp	kW
Horse			
galloping at 27 mph (12 m/sec)[a]	2 min	2	1.5
towing barge at 2.5 mph (1.1 m/sec)[b]	10 h	0.67	0.5
Man			
towing barge at 1.5–3 mph (0.67–1.34 m/sec)[b]	10 h	0.11	0.08
turning winch[b]	10 h	0.058	0.043
working treadmill[b]	10 h	0.081	0.06
climbing staircase[c]	8 h	0.12	0.09
turning winch[c]	2 min	0.51	0.38

Sources:
a. A. F. Burstall, *A History of Mechanical Engineering* (London: Faber and Faber, 1963).
b. R. D'Acres, *The Art of Water-Drawing* (London: Henry Brome, 1659 / Cambridge: Heffer, 1930).
c. reference 1.

output rather less than half the breathing capacity of an average man is involved, and informed opinion now suggests that this exertion is the maximum which could be expected without adverse effects on health for average men working for long periods.

Information on the energy cost of locomotion of animals other than man can be found in references 15–18.

In a review of the energy used per ton-mile (or tonne-km) and passenger mile (km) for such varied means of transportation as the S.S. *Queen Mary*, the supersonic transport, a rapid-transit system, and oil pipelines (ref. 8), Rice points out that a bicycle and rider are by far the most efficient. He calculates that a modest effort by a bicyclist which results in 72 miles (116 km) being covered in 6 hours could require an expenditure of about 1,800 kcal (7.54 MJ), which is in agreement with figure 7.5 for something between a roadster and a sports bicycle. Assuming a weight of 200 lb (90.6 kg) for rider and machine, Rice states that this figure is equivalent to 100 ton-miles (146 tonne-km) (or over 1,000 passenger-miles) per gallon (3.785 liters) of equivalent fuel. The *Queen Mary* managed, by contrast, 3–4 passenger-miles per gallon (1.27–1.70 passenger-km per liter).

The energy consumption of other modes in comparison with that of a bicyclist is shown in figure 7.12.

Figure 7.12
Energy cost of human movement and propulsion of various vehicles.

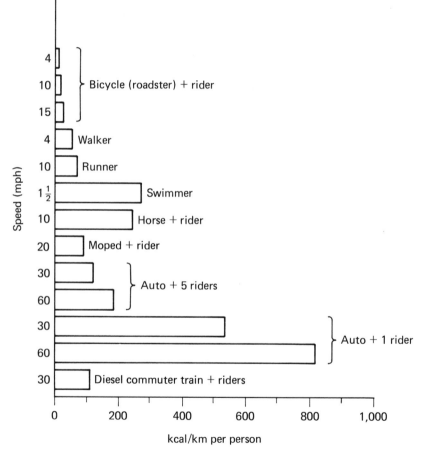

Bicycling versus other human-powered locomotion

Roller skating

From figure 7.1 it can be seen that for one hour of maximum power output the record speed credited to a roller skater (22.4 mph, or 10 m/sec) is less than that of a track bicyclist (30.7 mph, or 13.4 m/sec). If it is assumed that such record makers exert equal powers at their respective relative speeds, an estimate can be made of the rolling resistance of skates as follows. Assume that the skater has a frontal area of 3 ft² (2.079 m²), which is less than the 3.65 ft² (0.339 m²) of a very crouched bicyclist and his machine. At 22.4 mph (10 m/sec) a bicyclist exerts 0.25 hp to overcome air resistance (see figure 7.3). Therefore, the power needed by the skater to overcome air resistance is

$$(3/3.65) \times 0.25 \text{ hp} = 0.205 \text{ hp} (149 \text{ W}).$$

At 30.7 mph (13.7 m/sec), the bicyclist exerts 0.73 hp (544 W) (see figure 7.1), and we assume that the skater at 22.4 mph (10 m/sec) is exerting the same power. Hence, the power absorbed by the skates is

$$0.73 \text{ hp} - 0.1205 \text{ hp} = 0.525 \text{ hp} (392 \text{ W}).$$

If the skater weighs 154 lb (69.85 kg), the coefficient of rolling resistance of the skates is

$$\frac{392 \text{ W}}{69.85 \text{ kg} \times 9.81 \text{ m/sec}^2 \times 10 \text{ m/sec}} = 0.057.$$

The above rolling resistance is very high compared with that of bicycle wheels, assumed for the purposes of figure 7.3 as 0.003–0.008. The very large increase can be partly credited to the use of very small wheels in the skates (about $\frac{1}{13}$ the diameter of bicycle wheels) and to the less-easy running at high speed of the hard rollers compared with the pneumatic tires of the bicycle. Measurements of the pull required to keep a skater running steadily made by the senior author showed a rolling-resistance coefficient of about 0.060 at low speeds, and other information suggests that this would be greater at 21 mph (9.39 m/sec).

Several attempts are being made to produce skates having large wheels of much lower rolling resistance, to determine the effectiveness of this form of man-powered locomotion. Cross-country skiers train in summer on a form of large-wheeled roller skate (see figure 10.8).

Walking

For the purpose of comparison, table 2.1 has been drawn up from information given in references 19, 11, and 20 and elsewhere. The data of reference 19, which are the main source for table 2.1, can be interpreted as meaning that the maximum tractive resistance of the walker is about $\frac{1}{13}$ of his weight. This figure was given as early as 1860 (ref. 20). A higher resistance of 1/7.5 is, however, estimated from a simple geometrical model.[21]

The tables show that for the same breathing rate the bicyclist's speed is about four times that of the walker.

The metabolic-heat figures were obtained by multiplying the oxygen consumption, in liters per minute, by a calorific value constant of 5 kcal per liter of oxygen, given by Falls as a reasonable value for the circumstances.[22] This represents the total "burnup" of human tissue which must ultimately be replaced by food. If each kilocalorie could be converted in one minute at 100 percent efficiency to mechanical energy (via muscle action), 0.09 hp (69.8 W) should result.

Reference 19 shows that walking up a hill is slightly more efficient (in terms of energy consumption) than level walking, so the difference between cycling and walking is lessened in that case.

Running

The recorded times for sprint runners and racing bicyclists on level tracks in still air show that a cyclist can reach 40 mph (17.88 m/sec) for the furlong (220 yards, or 201.17 m) and 30 mph (13.41 m/sec) for the mile (1,609.3 m), whereas a runner reaches only half these speeds. Assuming that the wind resistance of a bicycle and rider and that of a runner are similar at similar speeds, we can estimate that the power needed for cycling is only about a fifth of that needed for running at the same speed, in the range of 15–20 mph (6.7–8.9 m/sec).

Effect of gradients and headwinds

Gradients and headwinds impede both the bicyclist and the walker, but to different degrees compared with level progression in still air. It can be calculated that a gradient of 4 percent (1/25) or a headwind of 10 mph (4.47 m/sec) slows a bicyclist exerting a constant 0.05 hp (37.3 W) to about 2.5 mph (1.12 m/sec). A walker developing the same power would be slowed from about 2 mph to 1.25 mph. The rider is slowed to 25 percent speed and the walker to about 55 percent. As a consequence, the rider notices difficult conditions more than the walker. On the other hand, with a tailwind or when going downhill the bicyclist is aided to a far greater extent than the walker, and it is probably this virtue of the bicycle that will ensure its use even in country with hills so steep that the bicycle must be pushed up them.

When a bicyclist or a walker climbs a hill his weight has to be lifted through a vertical distance, and as a consequence extra power is required above that needed for progress along the level. The additional power required for a bicycle and rider with a total weight of 170 lbf (756 N) to climb a hill of 5 percent (1/20) at 25 mph (11.18 m/sec) is

$$\frac{756 \text{ N} \times 11.18 \text{ m/sec}}{20} = 423 \text{ W}.$$

Hence, it is seen from figure 7.3 that a racing bicyclist climbing a 5-percent hill must exert a power of 0.57 + 0.407 or 0.97 hp (723.3 W). He would be sorely stressed and could do this for only about 2 minutes, according to figure 2.5.

Bradley gives interesting information about his climbing a $\frac{1}{12}$ (8.5-percent) pass on the Gross Glockner, of 12.5 miles (20.1 km) length, in about 57 minutes.[23] The gear used was 47 inches (3.76 m), and it can be deduced that he exerted at least 0.6 hp (447.6 W), pedaling at a rate of about 90 rpm. This performance is remarkably close to the fast 250 mile (40.2-km) time-trial performances listed in table 7.1, and provides convincing proof that there is sound evidence for all the power-requirement estimates based on wind-resistance calculations (as distinct from the more easily accepted simple weight-raising calculations associated with hill-climbing bicyclists).

Should one walk or pedal up hills?

Noncompetitive bicyclists have the option of walking up steep hills. Some prefer to do so, alleging that a change of muscle action is agreeable to them. Some bicyclists, however, prefer to fit low gears to their bicycles and to ride as much as possible. Whether it is easier to ride or to walk up steep gradients is often debated among bicyclists. We will use data developed previously to show that it should be more efficient to ride up to an approximately limiting gradient.

If we confine attention to the everyday bicyclist, we can assume that he is unlikely to wish to use more than about 0.1 hp (74.6 W). A commonly encountered steep hill is one with a gradient of 1/6.7, or 15 percent. It is assumed that the road speed which is thereby fixed as 1.5 mph (0.67 m/sec) gives no difficulties in balancing.

There have been many experiments on the oxygen consumption of pedalers.[24,25] The data given in figure 2.9 appear typical in that, for a power output of 0.1 hp (74.6 W) at the wheel, a metabolic gross efficiency of 21 percent is reasonable. The cyclist will be "lifting" a machine weighing, say, 30 lbf (130 N) in addition to his

body (150 lbf, or 667 N), so a factor is necessary
for the efficiency when compared with body
weight alone. This can be calculated as

21×150 lbf/$(150 + 30)$ lbf = 17.5 percent,

if one assumes that there is negligible rolling or
wind resistance at 1.5 mph and if one neglects
power losses in the low gear.

Reference 26 gives a summary of experimental
work concerning the oxygen consumption of
walkers going up various gradients at various
speeds. For a walking rate of 1.5 mph up a
grade of 15 percent, it appears that a metabolic
gross efficiency of 15 percent is accepted as typical. This efficiency assumes as a basis the body
weight being lifted against gravity. The bicyclist
pushing his machine will be in a semicrouched
position, so an adjustment to the efficiency must
be made. Data from references 19 and 26 concerning the effects of walking in stooped positions and when carrying small weights show
that pushing the 30-lbf bicycle absorbs 30 percent extra effort, so that the walker's muscle efficiency based on his body weight alone is
decreased to $17.5 \times (100 - 30)/100 = 12.3$ percent. From the estimations above, it appears that
it is easier to ride up a 15-percent gradient than
to walk at the same speed of 1.5 mph (0.67
m/sec), pushing the bicycle, by about 12.3/17.5,
or about 30 percent.

However, in practice, the lowest gear available
may be 20 inches (1.6 m), which gives a pedaling rate of 26 rpm—not optimal, according to
figures 2.3 and 2.9. A lowering of the previously
assumed overall pedaling efficiency of 21 percent is bound to occur. Let us estimate this at
about 18 percent. As a consequence, the 30-per-
cent difference quoted above should be taken as
about 18 percent. This difference gives only a
small margin for the extra transmission friction
involved in the use of a very low gear. Calculations along the lines of the above show that the
15-percent gradient may be a critical one, and

that at gradients of 20 percent there is no really appreciable advantage in riding the bicycle, even in a low gear.

A matter not given prominence in this type of discussion is the lack of wind cooling for the cyclist's relatively high heat output. At a power output of 0.11 hp or 82 W (that is, 0.1 hp plus an allowance for low-gear friction), a rider on the level would be traveling at some 14 mph (6.2 m/sec) and would receive considerable cooling. When climbing a hill at 1.5 mph for, say, 15 minutes, it is certain that an averagely clothed bicyclist would feel himself getting hot. Unpublished data suggest a body-temperature rise of appreciable magnitude: 1°F (0.55°C). It is probable that such considerations influence bicyclists to get off and walk at very low speeds (say, less than 1 mph) when the lower heat loss from the lowered power output is more tolerable. Proponents of very low gears for hill climbing can claim not only a higher metabolic efficiency but also a much-needed heat-removal effect from the more rapid movement of the legs at low forward speeds.

Human power versus engines and motors

Only two types of small power units have been developed for propelling light bicycles. The small internal-combustion engines used on mopeds and as auxiliary units on bicycles are well known. The other type is an electric motor that runs on lead-acid storage batteries; a scooter for use in factories uses this system. A Humber racing tandem of the 1890s (figure 7.13) was fitted with an electric motor and a frameful of batteries.

Specifications for modern mopeds show that a gasoline engine and accessories of power and endurance equivalent to that of a human pedaler would weigh about 20 lbm (9.07 kg). Performance details for electric propulsion show that this method would add about the weight of a man in the form of batteries and a motor with an output of about 0.1 hp.

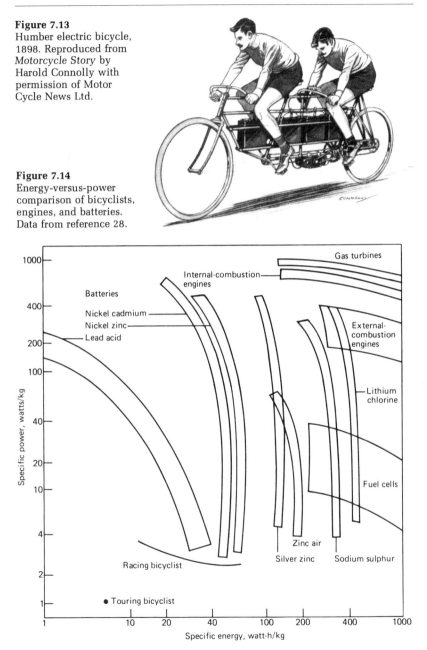

Figure 7.13
Humber electric bicycle,
1898. Reproduced from
Motorcycle Story by
Harold Connolly with
permission of Motor
Cycle News Ltd.

Figure 7.14
Energy-versus-power
comparison of bicyclists,
engines, and batteries.
Data from reference 28.

Reference 27 gives a chart (figure 7.14) of general data concerning the specific powers and energies of various power sources. We have added data for a racing cyclist riding at 20 mph for 24 hours, at 22 mph for 12 hours, and at 25 mph for 4 hours. The dot represents a touring cyclist covering about 100 miles in 8 hours. There is a degree of similarity between human energy capacity and that lead-acid batteries, even though pedalers' performances are not strictly comparable to those of batteries because the batteries are not recharged as the long-distance cyclist is through periodic snacks.

Other data on various heat engines and human performances are given in table 7.6.

Table 7.6 Energy cost of movement by various means.

	Speed		Energy consumption[a] per person		
	mph	m/sec	kcal/km	mpg[b]	km/l
Bicycle (roadster) plus rider	4	1.79	8.4	2440	1037
	10	4.47	15.6	1310	557
	15	6.70	24.4	840	357
Walker	4	1.79	55.3	370	157
Runner	10	4.47	68.3	300	127
Swimmer	$1\frac{1}{2}$	0.67	269.6	76.0	32.3
Horse + rider	10	4.47	245.4	83.5	35.5
Moped + rider	20	8.94	88.3	232	98.6
Auto + 5 riders	30	13.4	120.5	170	72.3
	60	26.8	183.0	112	47.6
Auto + 1 rider	30	13.4	539	38	16.2
	60	26.8	820	25	10.6
Diesel commuter train + riders	30	13.4	112	183	77.8

a. For the metabolic energies, these figures give the incremental consumption above the resting level.
b. Equivalent miles per U.S. gallon of 33,000-kcal/gal fuel (gasoline per person, calculated as follows: mpg = 33,000 × 0.621/kcal-km.

References

1. A. Sharp, *Bicycles and Tricycles* (London: Longmans, Green, 1896 / Cambridge, Mass.: MIT Press, 1977).

2. T. Nonweiler, Air Resistance of Racing Cyclists, report 106, College of Aeronautics, Cranfield, England, 1956.

3. A. Moulton, The Moulton Bicycle, Friday evening discourse, Royal Institution, London, 23 February 1973.

4. G. Buckett, A Bicycle Dynamometer, undergraduate project, Keble College, Oxford, 1974.

5. C. R. Kyle et al., Factors Affecting the Speed of a Bicycle, California State University, Long Beach, 2 November 1973; see also *Bicycling* (July 1974): 22–24.

6. F. R. Whitt, A note on the estimation of the energy expenditure of sporting cyclists, *Ergonomics* 14 (1971), no. 3: 419–424.

7. F. DeLong, *Guide to Bicycles and Bicycling* (Radnor, Pa.: Chilton, 1974).

8. L. G. C. E. Pugh, The relation of oxygen intake and speed in competition cycling and comparative observations of the bicycle ergometer, *Journal of Physiology* 141 (1974): 795–808.

9. E. J. Hermans-Telvy and R. A. Binkhorst, Lopen of fietson?—Kiesen op bass van het energieverbruick, *Hart Bulletin* (6 June 1974): 59–63.

10. J. D. Brooke and C. J. Davies, Comment on the estimations of energy expenditure of sporting cyclists, *Ergonomics* 16 (1973), no. 2: 237–238.

11. M. G. Bekker, *Theory of Land Locomotion* (Ann Arbor: University of Michigan Press, 1952).

12. J. C. Trautwine, *The Civil Engineer's Reference Book*, 21st edition (Ithaca, N.Y.: Trautwine, 1937), pp. 685–687.

13. W. C. Adams, Influence of age, sex and body weight on the energy expenditure of bicycle riding, *Journal of Applied Physiology* 22 (1967): 539–545.

14. C. H. Wyndham et al., Inter- and intra-individual differences in energy expenditure and mechanical efficiency, *Ergonomics* 9 (1966), no. 1: 17–29.

15. C. L. M. Kerkhoven, Kenelly's law, *Work Study and Industrial Engineering* 16 (February 1963): 48–66.

16. K. Schmidt-Nielson, Locomotion: Energy cost of swimming, flying and running, *Science* 17 (21 July 1972): 222–228.

17. S. S. Wilson, Bicycle technology, *Scientific American* 228 (March 1973): 81–91.

18. R. A. Rice, System energy and future transportation, *Technology Review* 74 (January 1972): 31–48.

19. G. A. Dean, An analysis of the energy expenditure in level and grade walking, *Ergonomics* 8 (1965), no. 1: 31–47.

20. "Velox" (pseudonym), *Velocipedes, Bicycles and Tricycles: How to Make and Use Them* (London: Routledge, 1869).

21. "An Experienced Velocipedist," *The Velocipede* (London: J. Bruton Crane Court, 1869), pp. 5–6.

22. H. B. Falls, *Exercise Physiology* (New York: Academic, 1968).

23. B. Bradley, My Gross Glockner ride, *Cycling* (25 July 1957): 90.

24. Report of the Bicycle Production and Technical Institute, Japan, 1968.

25. S. Dickenson, The efficiency of bicycle pedaling as affected by speed and load, *Journal of Physiology* 67 (1929): 242–245.

26. I. McDonald, Statistical studies of recorded energy expenditures of man. II. Expenditures on walking related to age, weight, sex, height, speed and gradient, *Nutrition Abstracts and Reviews* 31 (July 1961): 739–762.

27. S. W. Gouse, Steam cars, *Science Journal* 6 (1970), no. 1: 50–56.

Recommended reading

M. Denny, Locomotion: The cost of gastropod crawling, *Science* 208 (13 June 1980): 1288–1290.

R. M. Alexander, *Animal Mechanics* (Seattle: University of Washington Press, 1968).

8 Braking

The friction of dry solid substances

Experiments have shown that when two surfaces are pressed together with a force F, there is a limiting value R of the frictional resistance to motion. This limiting value is a definite fraction of F, and the ratio R/F is called the coefficient of friction, μ. Therefore, $R = \mu F$. For dry, rigid, surfaces, μ is affected little by the area of the surfaces in contact or the magnitude of F.

When surfaces start to move in relation to one another, the coefficient of friction falls in value and is dependent on the speed of the relative movement. For steel wheels on steel rails the coefficient of friction can be 0.25 when stationary and 0.145 at a relative velocity of 40 mph (17.9 m/sec). Polishing the surfaces lowers the coefficient of friction (one cause of brake fade), as does wetting. The coefficients of metal-to-metal dry friction are about 0.2–0.4 (down to 0.08 when lubricated); for leather to metal they are 0.3–0.5. All these are for stationary conditions and decrease with movement. Brake-lining materials against cast iron or steel have a friction coefficient of about 0.7, and this value decreases less with movement than for other materials. Elastomers (rubbery materials) deform under load, which causes their friction to be highly variable. In contrast with the case of dry rigid surfaces, the friction of elastomers is affected by contact area, increasing with greater area. Thus, such measures as "dimpling" brake rims can be counterproductive. The friction of elastomers is at a maximum when the material is made to "creep" along a surface. As true sliding begins, the coefficient of friction falls, decreasing with increasing relative velocity.

The variability of friction with contact area and relative motion, coupled with the flexibility of brake mechanisms which can change the contact area as the load increases, often leads to a "stick-slip" sequence, which, occurring repeatedly and rapidly, gives rise to brake squeal.

Bicycle brakes

Two places where solid-surface friction occurs must be considered in normal bicycle braking: the brake surfaces and the road-to-wheel contact. (This excludes track bicycles, which are braked by resisting the motion of the pedals, the rear cog being fixed to the wheel hub without a freewheel.)

Five types of brakes have been fitted to regular bicycles for ordinary road use.

The *plunger brake* is used on some present-day children's bicycles and tricycles, and was used on early bicycles such as the ordinary or penny farthing and on pneumatic-tired safeties up to about 1900 (figure 8.1). Pulling a lever on the handlebars presses a metal shoe (sometimes rubber-faced) against the outer surface of the tire. These were and are used on solid and pneumatic tires; the performance is affected by the amount of grit taken up by the tire, which fortunately increases braking effectiveness and wears the metal shoe rather than the tire. Such brakes are very poor in wet weather because the tire is being continuously wetted.

Figure 8.1
Plunger brake on Thomas Humber's safety bicycle. Reproduced with permission from Nottingham Castle museum.

The *internal-expanding hub brake* is similar to
the hub brakes of motorcycles and cars, but it is
less resistant to water and therefore its performance varies in wet weather. Hub brakes used
to be popular on medium-weight "roadsters" in
the 1930s, but they lost favor. They have recently been reintroduced in an improved form
by T. I. Sturmey Archer Ltd. (figures 8.2, 8.3).

The *backpedaling or "coaster" hub brake*
brings multiple disks or cones together when
the crank rotation is reversed (figure 8.4). These
brakes operate in oil and are entirely unaffected
by weather conditions. They are very effective
on the rear wheel; they cannot be fitted to the
front wheel because the actuating force required
is too great to be applied by hand. They cannot
be used with derailleur gears, and if the chain
breaks or comes off the sprockets there is no
braking at all.

The *disk brake* has recently been introduced
for bicycles. It is cable-operated from normal
hand levers (figure 8.5). The effective braking
diameter is at less than half the wheel diameter,
which requires a high braking force but keeps
the surfaces away from the wheel spray in wet
weather. These brakes are reputed to be effective in wet and dry weather.

The *rim brake* is the most popular type. A pad,
usually of rubber-composition material, is
forced against the inner or the side surfaces of
the wheel rims, front and rear. Because the
braking torque does not have to be transmitted
through the hub and spokes, as with the preceding three types, and because the braking force is
applied at a large radius, these brakes are intrinsically the lightest types and result in the
lightest bicycle design. Rim brakes are, however, very sensitive to water (the coefficient of
friction with regular combinations of brake
blocks and wheel materials has been found to
fall when wet to a tenth of the dry value[1]) and
to wear by the rim. Some blocks wear rapidly,
requiring continual adjustment (provided auto-

Figure 8.2
Sturmey-Archer internal-
expanding hub brake.
Courtesy of T. I.
Sturmey-Archer, Ltd.

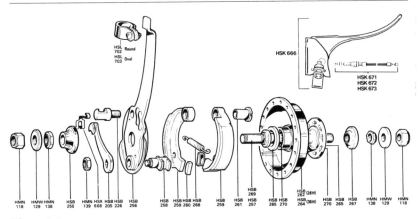

Figure 8.3
Exploded view of
Sturmey-Archer internal-
expanding hub brake.
Courtesy of T. I.
Sturmey-Archer, Ltd.

Figure 8.4
Exploded view of Bendix
backpedaling hub brake.
Courtesy of Bendix Corp.
Power and Engine
Components Group,
Elmira, N.Y.

Figure 8.5
Disk brake on rear wheel.
Courtesy of Phil Wood.

matically in some designs) and block replacement about every 2,000 miles (3,218 km).
Automobile brake shoes, with heavier duty,
last around 50,000 miles (80,500 km).

Duty of brake surfaces

Drum brakes for modern motor vehicles can be
designed by allowing a certain horsepower
(from 6 to 10) to be absorbed per square inch
(about 7–12 MW/m²) of braking surface.[2] The
power to be absorbed depends upon the speed
and mass of the vehicle and on the desired deceleration rate.

For a typical bicycle of 30 lb (13.6 kg) and
rider of 170 lb (77.1 kg), let us determine the
power loading at the brake blocks (assumed to
have a total area of 4 in.² or 2,581 mm²) if a
retardation of −0.5g (half gravitational acceleration) from 20 mph (8.94 m/sec) is required.
Gravitational acceleration, g, is 32.17 ft/sec²
(9.81 m/sec²), and expressing braking decelera-
tions as proportions of g is useful because it
gives directly the proportion of the vehicle's

and/or rider's weight which must be applied as braking force. The time t for a retardation a is given by

$$v_2 = v_1 + at,$$

where $v_2 = 0$ and v_1 is the initial velocity. Therefore, $v_1 = -at$, and so

$$t = -\frac{v_1}{a} = -\frac{8.94 \text{ m/sec}}{-0.5 \times 9.81 \text{ m/sec}^2} = 1.823 \text{ sec.}$$

The stopping distance is

$$S = \frac{v_1 + v_2}{2} t = 8.94 \frac{1.823}{2} = 8.15 \text{ m (26.7 ft)}$$

The initial kinetic energy is

$$\frac{mv^2}{2g_c} = \frac{(77.1 + 13.6) \text{ kg}}{2} (8.94 \text{ m/sec})^2$$

$$= 3{,}625 \text{ joule (2,672 ft-lbf).}$$

The power dissipation falls from a peak at initial application of the brakes to zero when the bicycle comes to rest. Determining brake duty—largely a function of surface heating—requires the mean power dissipation, KE/t, which is given by

$$\frac{3{,}625 \text{ J}}{1.823 \text{ sec}} = 1{,}988 \text{ W (2.67 hp).}$$

Thus, the power absorbed per unit of brake-block area is

$$\frac{1{,}988 \text{ W}}{2.581 \times 10_z{}^6 \text{ m}^2} = 0.77 \text{ MW/m}^2 \text{ (0.667 hp/in.}^2\text{).}$$

This is less than one-tenth of the average loading allowed in automobile-brake practice. Therefore, the surface area is more than adequate for braking. However, many riders in mountainous country have learned, to their dismay, that the thermal mass of and the heat transfer from a wheel rim are small. Rim brakes can cause the rim's temperature to rise quickly to the point at which the rubber cement holding tire patches, or even the tire itself, softens, and the tires will deflate or (in the case of "stick on" tubular tires) come off the rim. When these failures occur at speed on the front wheel, serious accidents are possible.

The adequacy of a vehicle's braking surface is, of course, only one factor in determining the distance in which the vehicle can be stopped. It is necessary in addition to be able to apply an adequate force to the brake system. Bicycle brakes are often deficient in this respect, especially in wet weather (when the coefficient of friction is greatly reduced) and especially for the front wheel (where most of the braking capacity is available).

Friction between tire and road

If we assume that an appropriate force can be applied to the brakes and that the blocks or linings have been proportioned so that they will not fade on account of heating, the stopping capacity of the brakes depends directly upon the grip (or coefficient of friction) of the tires on the road. For pneumatic-tired vehicles, this grip varies from 0.8 to 0.1 times the force between tire and road, according to whether the surface is dry concrete or wet ice.

Longitudinal stability during braking

The weight of the bicycle and rider does not divide itself equally between the two wheels, particularly during strong braking. To determine whether or not the braking reaction is important, let us estimate the changes in wheel reactions for the typical bicycle and rider above for braking at half the acceleration of gravity.

If the wheelbase is 42 inches (1.067 m) and the center of gravity of rider and machine is 17 inches (0.43 m) in front of the rear-wheel center and 45 inches (1.143 m) above the ground (figure 8.6), we can calculate the front-wheel reaction R_f when stationary or when riding at constant speed by equating moments about point 1 in figure 8.6:

Figure 8.6
Assumed configuration
for braking calculations.

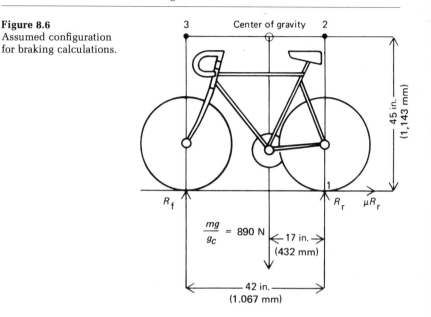

$R_f \times 1.067$ mm $= 90.7$ kg $\times 9.81$ m/sec^2
$\times 432$ mm,

so

$R_f = 360$ N $(81$ lbf$)$;

therefore,

$R_r = 890 - 360 = 530$ N $(119$ lbf$)$.

During the 0.5g braking, a total braking force of
$0.5 \times 890 = 445$ N $(100$ lbf$)$ acts along the road
surface. The front-wheel reaction R_f around
point 2 in the figure is now

$R_f \times 1{,}067$ mm $= 890$ N $\times 432$ mm
$+ 445$N $\times 1{,}143$ mm,

so

$R_f = 837$ N $(188.1$ lbf$)$;

by subtraction,

$R_r = 53$ N $(11.9$ lbf$)$.

Figure 8.7
MIT test setup for brake-block materials. The spring allows the test block to follow an inevitably uneven rim without large variations in force. Strain gauges in the support allow measurement of normal and tangential forces. Courtesy of Allen Armstrong, Positech, Inc.

Thus, the rear wheel is in only light contact with the ground. Only a slight pressure on the rear brake will cause the rear wheel to lock and skid. The front brake has to provide over 90 percent of the total retarding force at a deceleration of 0.5g, even if the tire-to-road coefficient of friction is at the high level of 0.8. Therefore, brakes that operate on the rear wheel only, however reliable and effective in themselves, are wholly insufficient to take care of emergencies.

Another conclusion from this calculation is that a deceleration of 0.5g (4.91 m/sec^2, or 11 mph/sec) is almost the maximum that can be risked by a crouched rider on level ground before he goes over the handlebars. We can calculate the maximum possible deceleration as a proportion P of g by setting $R_r = 0$ in the above case. Then, taking moments of force (torques) around point 3, we have

$$890 \text{ N} \times (1{,}067 - 432) \text{ mm} = P \times 890 \text{ N} \times 1{,}143 \text{ mm},$$

whence $P = 0.56g$, or 5.45 m/sec^2 (12.19 mph/sec). Tandem riders and car drivers do not have this limitation; if their brakes are adequate they can theoretically brake to the limit of tire-to-road adhesion. If the tire-to-road coefficient of friction is 0.8 they are theoretically capable of a deceleration of 0.8g, which is 60 percent greater than that of a bicyclist with the best possible brakes. For this reason—and many others—bicyclists should never "tailgate" motor vehicles.

Minimum braking distances for stable vehicles

If it is assumed that the slowing effect of air resistance is negligible, a relatively simple formula can be used to estimate the minimum stopping distance of a vehicle fitted with adequate braking capacity and having the center of gravity sufficiently low or rearward in relation to the wheelbase for there to be no danger of the rear wheels lifting (see ref. 2):

$$\text{Distance (m)} = \frac{[\text{Initial speed (m/sec)}]^2}{20(C_A + C_R)},$$

or

$$\text{Distance (ft)} = \frac{[\text{Initial speed (mph)}]^2}{30(C_A + C_R)},$$

where C_A is the coefficient of adhesion and C_R that of rolling resistance. (C_A is the value of the coefficient of friction, μ, of a rolling wheel just before skidding occurs.)

Table 8.1 gives typical values for the coefficients, and table 8.2 gives calculations for various speeds of pneumatic-tired vehicles and actual stopping distances for railway trains. In practice, greater distances are needed for braking than those based on the formula and on a high adhesion coefficient. The railway figures indicate that if an adhesion coefficient of 0.1 is assumed, the formula gives braking distances of about half those normally found in practice (ref. 2).

Table 8.1 Coefficients of adhesion and rolling resistance (motor car).

Surface	Coefficient of adhesion	Coefficient of rolling
Concrete or asphalt (dry)	0.8–0.9	0.014
Concrete or asphalt (wet)	0.4–0.7	0.014
Gravel, rolled	0.6–0.7	0.02
Sand, loose	0.3–0.4	0.14–0.3
Ice	0.1–0.2	0.014

Sources: Reference 2, p. 321; G. M. Carr and M. J. Ross, The MIRA single-wheel rolling resistance trailers, Motor Industries Research Association, Nuneaton, England, 1966.

Table 8.2 Stopping distances for bicycles, cars, and trains.

Speed (mph)	Stopping distance, pneumatic tires (ft)			Railway train, practical (ft)
	Calculated	Safety code, cycle	Safety code, car	
8	2.5	3		40
10	4			60
12	5.7	8		80
16	10	16		120
20	16	24	20	160
30	36		45	260
40	64		80	510
50	100		125	850
60	145		185	1,300

Note: The adhesion coefficient used for calculated stopping distances is 0.85. The other distances for pneumatic tires are quoted from Road Safety Codes. (The safety code is a set of guidelines published in Britain as minimum recommended standards.) All values are for stopping on dry concrete. Practical values for railway trains are included for comparative purposes.

Table 8.2 includes distances quoted in British road-safety codes[3] for best performance of pneumatic-tired vehicles. These are also about twice those calculated from the formula (with an assumed adhesion coefficient of a magnitude achievable under very good circumstances). The road-safety-code performance figures have been well checked by the Road Research Laboratory (U.K.), the 1963 report of which gives details of measurements carried out on "pedal cycles" of various types as well as many types of motor vehicles.[4] The braking distances listed for bicycles confirm the calculations made above, where it was found that a little better than 26 ft (8.14 m) was possible for stopping from 20 mph (8.94 m/sec) without overturning. If the rider sat well back over the rear wheel he would be able to shorten the distance a little further. However, evidence obtained from spot checking indicates that the average motor vehicle on the road needs about twice the quoted code distances for braking under specified conditions (ref. 4), and it may be assumed that the same "service factor" applies to bicycles.

Rear-wheel-only braking

Let us see what braking distance we may expect if the same rider and bicycle studied earlier, starting from 20 mph (8.94 m/sec), brake with the rear brake only to the limit of tire adhesion. We assume that the rear brake is strong enough to lock the wheel if desired, and that the coefficient of friction μ between the tire and the road surface is 0.8. Then the maximum retarding force is $0.8 \times R_r$, where R_r is the perpendicular reaction force at the rear wheel. This rear-wheel reaction force R_r is somewhat less than the value during steady level riding or when stationary, because the deceleration results in more reaction being taken by the front wheel. Let us take the moments of forces about point 3 in figure 8.6. Under the assumed static conditions the machine is in equilibrium:

$R_r \times 1{,}067$ mm $+ \mu R_r \times 1{,}143$ mm
$= 890$ N $\times (1{,}067 - 432)$ mm

with $\mu = 0.8$; thus,

$R_r = 285.2$ N (64.1 lbf).

Then the deceleration, a, as a ratio of gravitational acceleration, g, is given by Newton's law:

$$F = \frac{ma}{g_c},$$

$$a = \frac{Fg_c}{m} = \frac{-\mu R_r g_c}{m},$$

$$\frac{a}{g} = \frac{-\mu R_r g_c}{mg} = \frac{0.8 \times 285.2 \text{ N}}{890 \text{ N}} = 0.256;$$

$$a = -0.256g.$$

So the retardation is less than half the value at which, using the front brake to the maximum safe limit, the rider would be about to go over the handlebars (0.56g).

The time taken for this deceleration is given as before by

$$v_1 = -at,$$

$$t = \frac{-8.94 \text{ m/sec}}{-0.256 \times 9.81 \text{ m/sec}^2} = 3.56 \text{ sec}$$

and the stopping distance is given by

$$S = \frac{v_1 + v_2}{2} t = \frac{8.94 \text{ m/sec} \times 3.56 \text{ sec}}{2}$$

$$= 15.91 \text{ m} (52.2 \text{ ft}).$$

Therefore, the stopping distance is about twice that for reasonably safe front-wheel braking. In practice a longer stopping distance is likely, because a deceleration level sufficiently below the limit where skidding starts would be chosen.

Wet-weather braking   Wet conditions affect both road adhesion and the grip of rim brakes on the rim. Braking distances for bicycles equipped with conventional rim brakes are approximately quadrupled in wet weather (ref. 4). Cars, which are generally fitted with weatherproof disk or drum brakes, are not nearly as affected by wet weather.

Experiments using laboratory equipment to simulate wet-weather braking of a bicycle wheel (refs. 1 and 4) have yielded the following significant findings.

For brake blocks of normal size and composition running on a regular 26-inch (equivalent to 650 mm) plated steel wheel, tests at the Massachusetts Institute of Technology (see reference 1 and figures 8.7 and 8.8) showed that the wet coefficient of friction was less than a tenth of the dry value. Moreover, the wet wheel would

Figure 8.8
Friction coefficients for wet and dry braking. Rim materials: (○) nickel-chromium-plated steel, (▷) aluminum alloy. Data from reference 1 and from A. Armstrong, Dynamometer Tests of Brake-Pad Materials (report), Positech, Inc., ca. 1977.

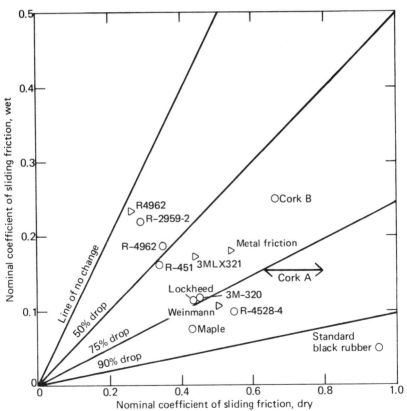

turn an average of 30 times with full brake pressure applied before the coefficient of friction began to rise, and a further 20 turns were necessary before the full dry coefficient of friction was attained (table 8.3). This recovery did not occur if water was being added to the brake blocks or rims after brake application, as might occur during actual riding in very wet conditions.

The measurement of coefficients of friction is frowned at by some investigators who believe that the notorious variability in measured values makes bicycle stopping distance from a standard speed of 15 mph (6.7 m/sec) on an actual or simulated bicycle the only valid measure. However, such stopping-distance tests have tended to confirm the validity of the MIT results.

Several different materials were investigated at MIT, and the results are shown in figure 8.8 and table 8.4. Although many of the materials are brake materials designated only by numbers, it can be seen that regular bicycle brake blocks ("B-rubber") have the highest dry coefficient and the lowest wet coefficient of friction of all materials tested. Attempts to improve the wet friction by cutting various grooves in the blocks or by using "dimpled" steel rims were unsuccessful. Similar findings have been reported by others.

The Road Research Laboratory found (ref. 4) that wet-weather performance can be improved by the use of brake blocks longer than the usual 2 inches (51 mm). Softer blocks than are common these days are also desirable, along with more rigidity in the brake mechanism and in the attachment to the frame of the brake itself.

Extra friction in the longer rear cable can decrease the force applied to the rear blocks by 20–60 percent compared with that at the front. Brake-cable "casings" with linings of low-friction plastic, such as PTFE, have been developed, and it is highly desirable that such

Table 8.3 Test data on operation of rim brakes; rims initially wetted.

Point	Braking force		Coefficient of friction,
	lbf	N	μ
1 (wet start)	22	97.9	0.17
2 (prerecovery)	22	97.9	0.17
3 (recovering)	26	115.7	0.20
4 (recovering)	31	137.9	0.24
5 (recovering)	35	155.7	0.27
6 (recovering)	39	173.5	0.30
7 (recovered)	44	195.7	0.34

Turns of wheel before onset of recovery	30
Turns of wheel during recovery	20
Total turns to recovery	50

Source: reference 1, page 32, run C-2.

Table 8.4 MIT data on brake-block materials for equivalent speed of 10 mph (4.5 m/sec).

Friction material	Nature of run	Average μ_{dry}	Average μ_{wet}	$\dfrac{\mu_{wet}}{\mu_{dry}}$	Turns to recovery	Remarks
R-451	dry	0.33	—	—	—	$\mu = 0.39$ at 120°F (48.9°C)
R-451	wet-dry	0.34	0.17	0.50	50	
B rubber	wet-dry	0.95	0.05	0.05	55	Erratic recovery
R-4528-4	wet-dry	0.55	0.10	0.18	54	
Maple	wet-dry	0.44	0.09	0.20	42	$\mu_{max} = 0.56$
Lockheed	wet-dry	0.45	0.12	0.27	25	during rec'y
R-451	wet-dry	0.34	0.17	0.50	53	
Cork A[a]	dry	0.63	—	0.42	—	
Cork A	wet	—	0.26			
Cork A	dry	0.79	—	0.24	—	
Cork A	wet	—	0.19			
Cork B[b]	dry	0.67	—	0.28		
Cork B	wet	—	0.19			
Cork A	wet[c]	—	0.16	—	—	
Cork B	wet[c]	—	0.25	—	—	
R-451	dry	0.43	—	—	—	
R-451	wet-dry	0.37	0.17	0.46	70	

Source: reference 1, p. 34.
a. Orientation A: Layers parallel to friction face.
b. Orientation B: Layers perpendicular to friction face.
c. After a 48-h soak.

casings become standard. However, it has been pointed out that the rear brake requires less actuating force than does the front if locking (skidding) is to be avoided. Although there have been commercializations of bicycle brakes with self-adjusting mechanisms, these were not successful. Virtually no present brakes allow adjustment without wrenches through the whole range of brake-block wear, a lack that leads to extremely dangerous conditions in bicycles ridden by less mechanically able persons.

The MIT tests were made with steel rims because of the severe dropoff in braking efficiency when rims of this material were used with any of the brake blocks then (1971) available. Since that time there have been several developments in wet-weather braking. These have been spurred partly by the aim of the International Standards Organization (Technical Committee TC/149) and of the U.S. Consumer Product Safety Commission to formulate generally acceptable safety standards for the performance of bicycle brakes in wet weather.

One development has been the introduction of a number of disk brakes (figure 8.4), which retain most of their stopping power when wet. Hub brakes have also been reintroduced in an improved form (figure 8.2). These can also be fitted to front wheels, which is desirable. However, the weight penalty incurred by the hub and by the heavier wheel it requires will prevent hub brakes being adapted to lightweight bicycles.

Aluminum-alloy rims have been standardized by some manufacturers. Compared with steel, aluminum usually gives a lower coefficient of friction with a given brake-block material in the dry and a higher coefficient in the wet. As this results in a smaller change in performance when going from dry to wet than occurs with steel rims, the use of aluminum rims is clearly an improvement. In some cases, there have been attendant disadvantages. The softer aluminum tends to be machined out of the rim if a piece of grit gets under a brake block. Pieces of aluminum become embedded in the brake block and oxidize to aluminum oxide (a hard, abrasive material), and then the rim is machined more rapidly and the dry coefficient of friction can fall to a dangerously low level. The junior author has had the braking surfaces of two aluminum wheels separate suddenly, blown out by tire pressure after the rim had been machined to below a critical thickness by brake action. Such an occurrence could lock the front wheel, with disastrous consequences. A possible solution to these problems of aluminum rims lies in one or more of a number of improved brake blocks recently marketed. Some are specifically for aluminum rims; others give much-improved wet-weather performance on steel rims.

It has recently been recognized that leather, which was first used for bicycle brakes before the turn of the century because of its good wear resistance, coefficient of friction, and ability to conform to the profile of the rim, also possesses outstanding wet-braking properties when used against a chrome-plated surface. This is true for chrome-tanned leather, but not, apparently, for leather tanned by the older "vegetable" process. It gives a ratio of wet-to-dry friction of between 0.5 and 1.0, for reasons not fully understood but connected with the porosity between the fibers and their affinity for water. No other material, synthetic or natural, has been found that can reproduce these qualities and at the same time stand up to the abrasion and temperatures inherent in the duty of a brake lining. Fibrax Ltd. has brought out a brake block in a leather reputed to be from buffalo hide. It is reported to give outstanding performance, with wet stopping distance no more than 30 percent greater than the dry. In 1980, Fibrax introduced a leather block for use with aluminum rims.

A rubber block of conventional shape but with a thin insert of special leather on the rubbing

Figure 8.9
Sturmey-Archer leather-
composite brake block.
Courtesy of T. I.
Sturmey-Archer, Ltd.

face (figure 8.9) was introduced in 1981. The
leather is treated to ensure a good bond to the
rubber, which is molded onto it, and to prevent
degradation during the molding process. The
manufacturer claims that, as well as solving the
wet-braking problem in a practical and economical way, the composite block has a life several
times that of a conventional rubber one, and
that if the leather wears through the performance simply reverts to that of a normal rubber
block. This block is the subject of a patent application. The leather fibers are surprisingly
hard. If used with an aluminum alloy rim they
will abrade the surface, causing the leather to
clog and the rim to wear, so this type of block
should only be used with steel rims. Raleigh
recommends that alloy rims should continue to
be used with high-quality rubber blocks, which
give a reasonable wet-braking performance; the
ratio of the wet to the dry coefficient of friction
is in the range 0.3–0.5.

Despite these seemingly excellent features,
leather brake blocks vary in performance and
cannot yet be said to be the complete answer to
the wet-braking problem.

A commercial elastomer has been produced as
blocks and as bonded brake shoes by the
Scott/Mathauser Corporation. We do not have
data on the coefficient of friction, but stopping-
distance tests have shown these blocks to give
the shortest distances of all brakes tested when
dry and to be second only to leather in the wet.
The material has very good wear resistance. The
excellent dry performance implies that users
must take care to avoid overbraking on the front
wheel.

A consequence of the MIT work with steel
rims was the development of a brake that could
use aircraft brake-pad materials found by Hanson to suffer very little drop in friction coefficient in going from dry to wet conditions. The
friction coefficient was too low to be used in a
regular caliper brake, because too large a

squeeze force would be required. It was not possible to strengthen a regular caliper brake and then to increase the leverage, because a consequence of increased leverage is decreased brake-pad motion. (Because bicycle wheels of present construction cannot be relied upon to run true, a considerable brake-pad gap must be allowed.) Therefore, a brake with two leverages was developed. When the brake lever is initially squeezed, the pads are moved under very low leverage (low force, large movement). As soon as the pads contact the rim, a slider in the brake mechanism locks up, and further movement has to take place through a high-leverage, high-force action.[5] The brake therefore has the additional advantage that it automatically takes up pad wear without further adjustment.

The dual-leverage brake was redesigned by Positech, Inc. (figure 8.10) and tested. Used on the front wheel only, with a regular caliper brake on the rear, it regularly achieved stopping distances of less than 25 percent of those given by regular brakes in wet conditions (3.5 m from 6.7 m/sec, instead of the usual 15–20 m). However, it has not been taken up commercially.

Occasionally, brakes are developed in which the braking forces themselves supply part of the

Figure 8.10
Positech dual-leverage brakes. The cylindrical body contains a slider that locks when the pads contact the rim. Subsequent movement of the actuating cable causes pad movement through the large leverage about the caliper pivot instead of through the small leverage through the chain link.

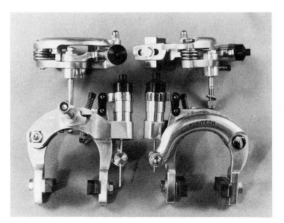

Figure 8.11
Servo-action brake
blocks. Arrow indicates
direction of rim motion.

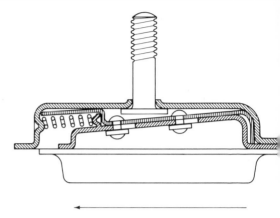

actuating force ("servo-action" brakes). A recently marketed French system (figure 8.11) incorporates angled ramps within the brake shoes,
so that the brakes, in being pulled forward by
the wheel rim during braking, are also forced
inward to give a stronger squeeze (but only if
there is significant friction in the cable, so that
the hand lever is not merely pushed out). A disadvantage of such "positive-feedback" arrangements is that they magnify the differences
between dry and wet friction coefficients. The
brake may give strong braking action with a
light actuating force when dry but provide insufficient braking even with a maximum
squeeze action when wet. What is needed,
rather, is an added "negative feedback" stage to
limit braking in dry conditions to below the
amount that would result in the rider being
projected over the handlebars. A braking system
incorporating such a combination of positive
and negative feedback (figure 8.12) has been developed by Calderazzo (R. C. Hopgood, personal
communication to D.G.W., 19 January 1979).
Only the rear-wheel brake is actuated by the
rider. The brake is mounted on a lever pivoted
near the wheel axis so that it is carried forward
during braking. In doing so, it actuates (through
a cable) the front brake, with any reasonable de-

Figure 8.12
Calderazzo feedback
brake system. When
handbrake is operated,
rear brake is carried
forward on slider against
spring, actuating front
brake simultaneously. If
bicycle starts to pitch
forward, rear wheel is no
longer rotated by road
surface, and front brake
is released.

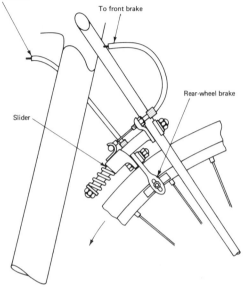

sired degree of force multiplication. Accordingly, little effort need be required for strong braking to be obtained. As soon as the degree of braking that causes the rear wheel to start skidding is reached, the braking at the front wheel is automatically limited. In hundreds of tests with this system, in which testers made "panic stops" from high speeds on different surfaces, never did a rider even begin to go over the handlebars. (The front forks of the test bicycle eventually failed through fatigue—testimony to braking effectiveness and fork-design inadequacy.) As of 1980, this promising system is tied up in patent litigation.

It has been found that, even when surfaces roll upon one another, a certain amount of slipping takes place, and that this leads to frictional losses. This phenomenon is rooted in the fact that the surfaces, however "hard," create cavities at the places of contact, and that these lead to alternate compression and expansion of the materials at these points and, as a consequence, to expenditure of energy.[6] With soft surfaces, the effects are pronounced but are well worth putting up with where vehicle tires are concerned because of the comfortable ride.

Although efficient tread patterns are essential for the good grip of automobile tires on the road at high speeds under wet conditions, it appears that at bicycle speeds the requirements for bicycle tires are not so stringent. Data from some tests suggest that no appreciable variation in the grip of a tire on the road under wet conditions could be expected from any design alteration (ref. 4). Below 20 mph (8.94 m/sec), nearly smooth patterns should suffice. This prediction is verified by the only slightly corrugated surfaces used for years on racing-bicycle tires.

Backpedaling

As stated above, track bicycles are braked by backpedaling. The idea that the rider should perform work to "destroy" energy has intrigued many people since the early days of bicycling. Horse-drawn vehicles have been braked in this way for thousands of years, and people running down stairs and steep slopes experience a similar muscle action.

Much discussion was devoted in the past to comparing the muscle actions used in forward and backward pedaling. Sharp concluded that muscle physiology played an equal part with mechanical motion.[7] He devised the interesting chart shown in figure 8.13 in the course of his writings on this subject. Time has proved Sharp's surmise correct, in that research has shown that for a given oxygen consumption a pedaler can resist power supplied by an animate or an inanimate prime mover more efficiently than he can perform ordinary forward pedaling.[8,9] A classic experiment in which a forward ergometer pedaler is resisted by a backward pedaler demonstrated vividly this difference in energy cost. The basic physiological reasons, involving muscle-action theory, are still being debated in the literature under the heading of "negative" or "eccentric" work.

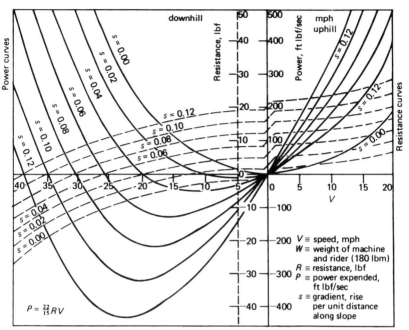

$R = Ws \pm (0.01W + 0.013V^2)$

Figure 8.13
Power expended in backpedaling. Dashed lines are resistance curves representing rolling plus aerodynamic drag. Solid lines are power curves. s is the gradient expressed as percentage/100 (for example, 0.12 is 1 in 8.5). Intercepts of power curves with horizontal axis show terminal downhill speed for each gradient. Between these velocities and zero velicity the "negative" power that has to be exerted in backpedaling goes through a maximum. From reference 7.

References

1. B. D. Hanson, Wet-Weather-Effective Bicycle Rim Brake: An Exercise in Product Development, M.S. thesis, Massachusetts Institute of Technology, 1971.

2. *Kempe's Engineer's Year Book*, vol. 11 (London: Morgan, 1962), pp. 320, 353.

3. *Safe Cycling* (London: HMSO, 1957).

4. *Research on Road Safety* (London: HMSO, 1963).

5. F. DeLong, The Positech brake, *Bicycling* (Emmaus, Pa.) (November 1976): 38–39.

6. *A Dictionary of Applied Physics*, ed. R. Glazebrook (London: Macmillan, 1922).

7. A. Sharp, Back-pedalling and muscular action, *CTC Gazette* (1899): 500–501.

8. B. C. Abbott, B. Bigland, and J. M. Ritchie, The physiological cost of negative work, *Journal of Physiology* 117 (1952): 380–390.

9. H. B. Falls, *Exercise Physiology* (New York: Academic, 1968), pp. 292–294.

Recommended reading

D. Bianca, Sliding Friction and Abrasion of Elastomers, report, Canadian Institute of Mining Metallurgy, 1967; reviewed in *Engineering* (London) (4 August 1967): 182.

9 Balancing and steering

The balancing and steering of bicycles is an extremely complex subject on which there is a great deal of experience and rather little science. We will report what we believe to be the best of each. Both approaches—experience and science—have attempted to answer two related but different questions:

- What are the geometrical relationships that can give the single-track vehicle, considered as a rigid body, "good" steering characteristics? This question concerns riding at any speed, but particularly at low speed, when steering angles (the angle through which the handlebar is moved by the rider) can be large. Above 2–3 m/sec (9–13 mph) the handlebar cannot be turned by more than a few degrees to either side of the straight-ahead (neutral) position without the rider being thrown out of balance and, usually, off the bicycle.

- What other factors, when combined with geometrical relationships, can avoid the steering instabilities known as shimmy (a rapid oscillation of the front wheel about the neutral position, usually occurring rather suddenly at fairly high speed)?

A vibration or an oscillation is usually similar to the bouncing of a weight hung on an elastic thread. The occurrence of steering oscillations implies, therefore, that the elasticity of the structure and possibly of the rider is involved. Let us defer discussion of this complex question until after we have considered the still-complicated question of the steering characteristics of a bicycle considered as a rigid body.

Steering characteristics of nonflexing bicycles

The rigid-body geometry is complicated because of the many angles involved and because of the offset of the front fork (figures 9.1 and 9.2). Three important angles are

the steering-head angle (the most important of the angles defining the frame), which is usually between 68° and 75°,

the steering angle, or the angle of the handlebars from the straight-ahead, neutral position, and

the angle of lean of the bicycle frame to the horizontal.

If the fork had no offset, the front wheel would sweep out a sphere as it was turned. With the offset, the wheel sweeps out a "doughnut." It is the combination of bicycle lean with the front-wheel position as a slice of the doughnut that makes bicycle geometry so complex. An additional problem when we account for the gravitational and centripetal forces occurs if we wish

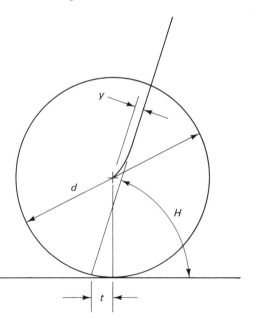

Figure 9.1
Front-fork geometry. H: head angle. y: fork offset. d: wheel diameter. t: trail.

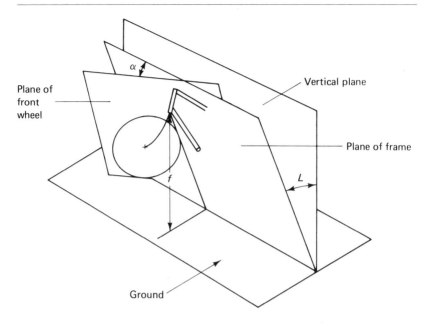

Figure 9.2
Steering geometry. α: steering angle. f: frame height. L: lean angle. Adapted from reference 7.

to account for a nonhorizontal ground surface.

There is no real disagreement about how a rider steers and balances a bicycle. One steers into or under a fall, just as one balances a broomstick on a finger. The following questions have intrigued many people, including some famous mathematicians and applied mechanicians such as Timoshenko (ref. 1):

Why are some bicycles easier to steer than others?

Why do some bicycles steer themselves easily, whereas others do not?

What are the effects of steering angle, fork offset or trail, height of center of gravity, and so forth?

However, members of the lay public who have perused the scientific literature (refs. 1–6) have been intrigued to find often complete disagreement among the experts, even about fundamentals. One advocates a high center of gravity for stability; another concludes from the equations that a low mass center is desirable. One finds that gyroscopic action is important; another the exact opposite. We have found that the most useful and relevant information about bicycle steering and stability is that given by David Jones, a chemist who looked into bicycle stability as a diversionary project.[7]

Jones set out to build an unridable bicycle (URB). In his URB I, he canceled out the gyroscopic action of the front wheel by mounting near it another similar wheel which he could rotate backwards. He found that this made little difference to normal handling, and concluded that gyroscopic action has little influence on bicycle stability. He did find, however, that URB I would not travel riderless. Gyroscopic action was important for the lightweight bicycle alone, but not for the bicycle plus rider, when the rider was controlling the bicycle with the handlebars. When Jones attempted to ride URB I "no hands," he could only just maintain his seat. The bicycle seemed to lack balance and responsiveness. This confirmed Den Hartog's analysis (ref. 2).

We will not relate in equal detail Jones's several other URBs, with large, small, and reversed fork offset and with a tiny front wheel. Suffice it to say that he was able to ride all of them, although URB IV, with a very large fork offset, was unstable and very difficult to ride. URB III had reversed fork offset and therefore a very large trail, and was extremely stable. When pushed, riderless, it would steer itself for an astonishingly long time, negotiating depressions and bumps in the road and continuing until it was almost stationary before falling over. It was, however, sluggish and heavy to steer on any path other than that dictated in some way by its interaction with the roadway. (D.G.W. had an old car with some of these characteristics; every bump or hollow in the road would change the direction without any movement of the steering wheel, and to maintain a straight course required continuous anticipatory steering. A high degree of stability is not always desirable in a vehicle.)

Jones quantified a stability function after making the following observations: When the bicycle is wheeled by, for instance, holding the saddle, it is caster action that makes the front wheel go straight ahead when the bicycle is moved with the frame vertical on a horizontal roadway. The front wheel "trails" the frame. The rear wheel trails along, too. One can immediately see the importance of trail by pulling the bicycle backward. If the steering-head bearings are free, the front wheel (now in the rear) will immediately flop around to some large steering angle. This movement of the front wheel is not at first assisted by gravity. The wheel sweeps out a doughnut, and regardless of the head angle (so long as it is less than 90°) or the fork offset (so long as it is finite) the front wheel is at a point of unstable equilibrium. That is, it requires a small disturbance to make it flop over. When the frame of the bicycle is tilted, the wheel is no longer in equilibrium in the straight-ahead position. There is a force or turning moment that increases the frame tilt and acts to turn the handlebars. The equilibrium angle taken up by the handlebars is, however, an inverse function of frame tilt. In other words, with a small tilt the handlebars turn a long way, and with a large tilt the handlebars go only a few degrees from the neutral position. The reason the handlebars turn is that this allows the frame to fall; the frame and the weight carried on it seek the minimum-potential-energy position.

The computer program Jones wrote to solve the steering geometry of a single-track vehicle with a rigid body and thin wheels (no allowance was made for tire cross-section shape) produced graphs like figure 9.3, which is for one steering-head angle and one fork offset (as a proportion of wheel diameter). He decided to

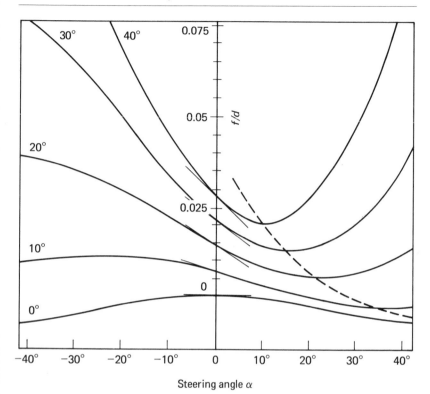

Figure 9.3
Typical results of stability calculations. Degree values on curves represent lean angle L; axis at middle represents relative frame height f/d; slopes indicated near middle are $[d\,(f/d)/d\alpha]_{\alpha=0}$. Head angle H is 70°; fork offset y/d is 0.094.

concentrate on the steering characteristics when the steering angle was near to zero, as in normal riding. Small changes in steering angle would produce a rise or fall in the frame height $(df/d\alpha)$. Jones reasoned that, for stable steering, the steering should "want" to turn into the curve as the frame leans around a bend. The reason for the steering wanting to turn had to be the fall of the frame in these circumstances. This criterion is expressed mathematically as the requirement that

$$\left(\frac{\partial^2(f/d)}{\partial\alpha\partial L}\right)_{\alpha=0}$$

be negative. From his computer program Jones produced figure 9.4, which covers all likely combinations of head angle and what he calls

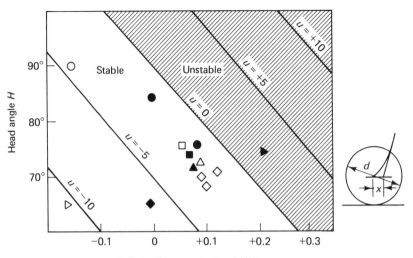

Relative front projection (*x*/*d*)

Figure 9.4
Jones's stability criterion.
Diagonal lines represent
constant stability, u ≡
[∂²(*f*/*d*)/∂α∂L]ₐ₌₀. (○)
1887 Rudge; (●) high-
wheelers; (□) modern
track bike; (■) modern
road bike; (△) modern
tourist bike; (▲) Raleigh
"chopper"; (◇) Ross
child's bike; (◆) 1879
Lawson safety; (▷) URB
III; (▶) URB IV. Note:
Because high-wheelers
differ from safeties in
that the rider pedals and
straddles the front wheel,
the points for high-
wheelers are included for
interest only.

"front projection," identified on figure 9.1. He
found that the graph agreed with experience.
URB IV was indeed in the unstable region,
while URB III was far into the stable region. We
have added some other bicycles to those Jones
considered.

The information in figure 9.4 can be expressed
in more familiar terms as

$$(y/d) = 0.00917[(90° - H)(\sin H) + 4u], \qquad (9.1)$$

where *y* is the fork offset, *d* is the wheel diameter, *H* is the head angle (in degrees), and *u* is
the stability criterion

$$\left(\frac{\partial^2(f/d)}{\partial\alpha\partial L}\right)_{\alpha=0}.$$

Experience indicates that bicycles have good
steering characteristics when u is between −1
and −3.

**Range of practicable
configurations for
standard bicycles**

The standard diamond-frame safety bicycle has
resulted in the universal selection of steering
angles from a very small range. This range has
emerged from the following considerations. The
crank length has generally been chosen at 170
mm to suit the majority of adult riders. The
height of the bottom bracket above the ground
has then been fixed so that the pedals clear the
ground in at least low-speed cornering. The rear
wheel is brought as close to the bottom bracket
as can reasonably be arranged, with the seat-
tube angle positioning the saddle so that the
rider's center of gravity is reasonably forward of
the rear-wheel center even when an upright riding position is used. (Otherwise the front wheel
would lift off the ground every time one attempted rapid acceleration.) Then, in touring or
commuting bicycles, the front wheel plus a possible fender or mudguard is brought as close to
the bottom bracket as possible without there
being a possibility of the feet or toeclips catch-
ing on the fender or fender stays during a turn.
If a large steering-head angle is used with this
proviso, the top tube or crossbar becomes long,
requiring a long reach to the handlebars. Accordingly, a relatively small head angle is used
for touring and commuting bicycles. Racing and
track bicycles do not use fenders, and the designers allow the possibility of interference between the toe clips and the tire because the skill
of the rider can be relied upon to avoid it.
Therefore, a relatively large steering-head angle
can be used on a racing bicycle, giving a
smaller and more rigid frame. The consequence
is that touring bicycles generally have head angles of 72°–73°, road-racing bicycles have angles
of 73°–74°, and track bicycles have angles of
74°–75°. We have calculated the stability index
u for some of the bicycles listed in reference 8
(see table 9.1). It is surprising, and gratifying, to
see the small range of u values used by designers. This seems to confirm the value of Jones's
work. As might be expected, there is a tendency

Table 9.1 Steering geometries and stability indices of high-quality bicycles.

Bicycle type	Head angle	Fork-offset ratio[a]	Stability index u[b]
Touring	72°	0.0736	−2.27
	72°	0.0740	−1.99
	72°	0.0692	−1.86
	73°	0.0845	−2.28
Road-racing	73°	0.0837	−2.26
	74°	0.0729	−2.00
	74°	0.0976	−2.64
	74.5°	0.0804	−2.20
Track	75°	0.0759	−2.09
	75°	0.0953	−2.60

a. Fork offset/Wheel diameter.
b. $u \equiv [\partial^2(f/d)/\partial\alpha\partial L]_{\alpha=0}$.

for the high-speed road-racing and track machines to have u values in the more stable range (from −2.0 to −2.65) and for the touring machines to use u values from −1.85 to −2.3, which give somewhat lighter, more responsive steering but still give plenty of stability according to Jones's criterion. The overlap between the touring and the racing machines is notable. It would seem that one could specify u = −2.0 for any type of bicycle and simply specify the head angle at 72°–73° for touring or 73°–75° for racing. Any variations of u from −2.0 would be for personal taste rather than because of any safety considerations. However, we have given a wider range of u (from −1.0 to −3.0) in table 9.2, together with a range of head angle from 70° to 76°, so that the fork offset may be specified by interpolation if desired.

Equation 9.1 may also be used. For example, if we wish to specify the fork offset y for a track bicycle with wheels 680 mm in diameter and a head angle of 74°, and if we choose u = −2.25, the equation gives

$y = 680 \times 0.00917[(90 - 74) \sin 74 - 4 \times 2.25]$
$= 39.78$ mm (1.565 in.).

The trail, t, is also given in table 9.1. It may also be calculated from the formula

$$\frac{t}{d} = \frac{1}{\sin H} \left(\frac{\cos H}{2} - \frac{y}{d} \right).$$

However, trail is a dependent variable, and not of primary importance. The frame builder works to a fork offset, y, and this should be specified from the head angle H and the desired stability value u, using table 9.2 or the formula upon which it was based.

Table 9.2 Ratio of fork offset to wheel diameter for various stability indices.

Head angle	Stability index u		
	-1.0	-2.0	-3.0
70°	0.135	0.099	0.062
72°	0.120	0.083	0.047
74°	0.104	0.067	0.031
76°	0.088	0.051	0.015

Shimmy

This phenomenon is well but trivially illustrated by many small carts, such as those often used in food markets, whose castered wheels oscillate through a large angle when the cart is pushed above a certain critical speed. Shimmy is dangerous when it occurs in vehicles carrying people. When airplanes switched from having a single trailing tail wheel to a single leading nose wheel (which, of course, was mounted with a degree of trail), many lives and planes were lost when a nose-wheel would suddenly shimmy to the point where control was lost or some part of the structure failed.[9]

A shimmy-type oscillation occurs in a system with mass, structural springiness, and damping if a mechanism arises to reinforce a random initial oscillation and if the damping is small. By "damping" we mean some form of frictional dissipation. A ball bouncing on the end of a spring will have some damping in the spring material and more damping in the air resistance; if the ball is lowered into a pool of water, the additional damping stops the oscillation very quickly.

Bicycle front-wheel shimmy probably happens as follows: Something, a bump in the road perhaps, causes a sudden change in steering angle when the bicycle is going straight ahead. The machine is going too fast to respond by turning in the direction of steer. Rather, the inertia of the bicycle and rider carries them forward, and the caster action of the front wheel produces a very large restoring moment. Because of the mass and gyroscopic inertia of the front wheel, it does not respond exactly in phase with the restoring moment. Rather, some energy is stored in flexing of the forks, of the handlebar stem, and perhaps of the wheel itself. Most of this energy is converted to kinetic energy when the wheel passes through its neutral position, causing it to overshoot and to repeat the process. There is not much damping in this system so long as the oscillations are small. When they build up to the point where the handlebars move appreciably, much of the additional energy will be lost in the friction between the hands and the grips. There will also be increased losses in the tire "scrubbing" on the road. But the oscillations may still be large enough to cause loss of control.

Shimmy may be influenced by loads carried over the front wheel, or by looseness in various joints and bearings. There is no universal cure for shimmy. It should be helpful to increase the stiffness of all components, especially the front fork, the handlebars and stem, and perhaps especially the wheel. Spokes should be pulled up to produce a high stress. Paradoxically, this usually results in longer spoke life than if the stress is low enough to allow considerable spoke flexing. Increased stiffness will increase the natural frequency, which will increase the speed at which shimmy could occur and will reduce the amount of energy stored in the vibrations. The inherent damping (friction) in the tire-road contact and in the fork-frame-handlebar structure may then be sufficient to suppress shimmy altogether.

Other factors complicating steering

The analyses by Jones and others involved fairly drastic simplifying assumptions, while showing by experiment that in normal circumstances these simplifications were justifiable. Here we will list some of the real-life factors usually ignored, and then mention anecdotally some of the steering problems that have been difficult to analyze or even to experiment with.

Tire slip

When there is a side force on the wheels (such as when there is a sidewind, or when the bicycle is being ridden along a sloping surface or a curved path), the tires "slip" in the direction of the side force. The angle of the slip depends on the ratio between the side force and the normal force, on the angle between the plane of the wheel and the ground, on the tire pressure, and on the tire's construction. Typical graphical relations for the slip angle are shown in figure 9.5.

Rider's steering response

The rider responds to perceived changes in balance (for instance) by moving the handlebars. Each rider has a different response and a different delay before initiating the response (ref. 7).

Wheelbase

Short-wheelbase bicycles are said to be "responsive," whereas long-wheelbase cycles (such as tandems) are "sluggish."

Mass of bicycle

The mass of a bicycle and the location of the center of mass affect steering behavior.

Figure 9.5
Typical slip angles for
bicycle tires at various
inclinations. From
reference 6.

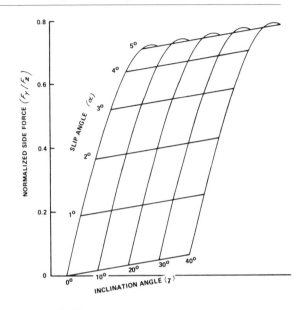

Mass of rider

The mass of the rider (more particularly, the relation of the rider's mass to the machine's) and
the relative position of the rider's center of mass
influence steering behavior.

Wheel moment of inertia

The higher the moment of inertia of the wheel,
the higher is the gyroscopic torque produced
when the plane of the wheel is turned.

Road-surface inclination angle

The angle the road surface makes with the horizontal significantly affects the steering forces
and the tire-slip angle (figure 9.5).

Angle of rider

Many riders try to hold themselves in the same
plane as the bicycle under all conditions. Others
hold their bodies at an angle to the plane of the
cycle, particularly when going around a curve.
In doing so they produce a different bicycle in-
clination angle than if the rider's center of mass

had remained in the plane of the bicycle, and the steering response is changed.

Rider-bicycle connection

A bicycle may be ridden with the feet in toe clips, the crotch firmly on an unsprung saddle, and the hands gripping the metal of the handlebars. Conversely, the rider may have a much looser or more flexible connection with the bicycle through a deeply sprung saddle, sponge-filled handlebar grips, and rubber-tread pedals, or he may ride with hands off the handlebars, crotch off the saddle, or feet off the pedals. In all of these circumstances the response of the machine will vary.

This list by no means exhausts the components of bicycle-riding characteristics—for instance, the springiness of the frame and the slack and friction in the steering bearings are of some importance. However, the above are probably the most important factors. Many of these factors are nonlinear. Mathematical analysis is understandably difficult in such a system; computer simulation is more appropriate. A simulation by Roland[10] has been the most comprehensive so far and has shown considerable success. An example of the simulation and of angles measured from an instrumented rider and bicycle is shown in figure 9.6. (Side-force loading was supplied by a small rocket motor—a hazard not often encountered on the road.) The computer was programmed not only to provide graphical responses, but also to illustrate the rider in an elementary manner as shown in figure 9.7.

Most of the conclusions of Roland's study would be acceptable by experienced bicyclists, and in fact would be regarded as common sense. This likely reaction gives one confidence in the method, and one wishes that it could have been carried farther into areas where we do not have good answers. Here, for instances, are two experiences of the junior author, one in-

Figure 9.6
Comparison of simulated
and experimental bicycle
responses after
disturbance of steering
torque. From reference
10.

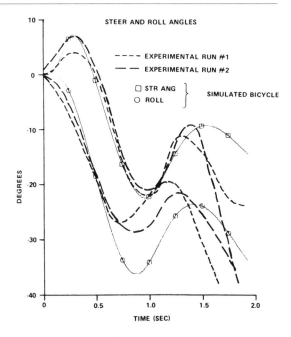

volving a potentially serious safety hazard and
the other more a comment on a tradeoff.

Of a fairly wide range of small-wheeled bicycles the author had ridden, one had a significantly safety hazard. It was a Moulton Speedsix
with a $16 \times 1\frac{1}{4}$ inch (406×28.6 mm) front tire
on an aluminum-alloy rim. On two occasions
when the front tire had sudden blowouts, the
steering became unmanageable. On the second
occasion the front wheel slipped sideways, taking the bicycle into the middle of the road in
front of a bus, which fortunately was able to
stop. Was this strange behavior attributable to
the steering geometry, or to some characteristic
of the tire and wheel when the tire was de-
flated? As this tire and wheel size is no longer
available, the question may be academic; no
similar characteristics have been found with
various bicycles with $16 \times 1\frac{3}{8}$ inch (406×34.9
mm) tires, despite experiments aimed at induc-
ing instability.

Figure 9.7
Computer-graphics
rendition of bicycle and
rider. From reference 10.

Figure 9.8
Avatar 1000 recumbent
bicycle.

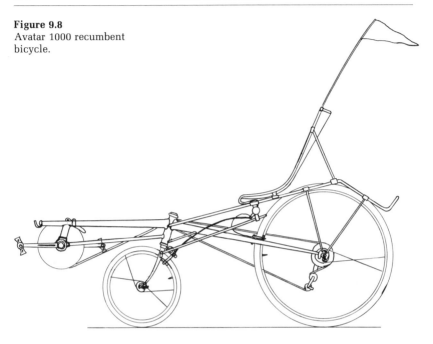

Figure 9.9
Avatar 2000 recumbent
bicycle. Courtesy of
FOMAC, Inc.

The second experience is related to the effect of the height of the center of mass of a bicycle and rider. D.G.W. has bicycled over several hundred thousand miles on regular diamond-frame sports-tourist bicycles and on Moultons, but for most of the past decade has used recumbent bicycles of short and long wheelbase (figures 9.8, 9.9). He finds little advantage or disadvantage of one configuration over another as regards steering characteristics in normal circumstances. Each novel configuration required a period of familiarization and then a longer period during which the rider's automatic responses were tuned to the particular characteristics of the type. However, there are some perhaps abnormal circumstances in which there is a sharp difference between the handling of machines with low and high centers of mass. These circumstances are those of an icy or snowy roadway, particularly when other vehicles have worn tracks in the ice or snow. In these conditions, a sideways skid of the front or rear wheel of a bicycle may be limited in extent. The wheel in effect may jump from one rut to another. For a given distance of side-slip, the resulting angle of lean is much greater for the low-mass-center than for the high-mass-center bicycle. The consequence is that many skids that would be quite tolerable on the higher machine cannot be escaped from on the low-mass-center bicycle. The tradeoff is that, although it might appear from the above that a regular (higher) bicycle is preferable to the recumbent, the danger of serious injury increases sharply with the height of the fall. One could take the illustration a step farther: a high-wheeler (preferably fitted with pneumatic tires) would probably be the best configuration for ice and snow, but the danger of serious injury in a fall from a high-wheeler was, and is, very great.

Alternative designs

Rear-wheel steering

Many people have seen theoretical advantages
in the facts that front-drive, rear-steering recumbent bicycles would have simpler transmissions
than rear-drive recumbents and could have the
center of mass nearer the front wheel than the
rear. The U.S. Department of Transportation
commissioned the construction of a safe motorcycle with this configuration. It turned out to be
safe in an unexpected way: No one could ride
it. The same is true of a rear-wheel-steering bicycle of the normal "safety" configuration constructed at the Bendix Company in Elmira, N.Y.,
which was reported to have stumped all comers.
L. H. Laiterman, in an undergraduate project at
the Massachusetts Institute of Technology,[11]
made a rear-steering recumbent in which all
significant aspects of the configuration could be
adjusted, and after considerable practice did
learn to ride the machine, albeit in an unsteady
manner.

Even if it were possible to further improve the
rear-steering configuration (whether of the recumbent design or not) and to learn how to
control such a machine fairly predictably, there
would be obvious disadvantages in, for instance, avoiding a suddenly opened car door by
first steering into the direction of the car with
the rear wheel. The essence of the unsteadiness
is related to this aspect of obstacle avoidance: as
one begins to fall to one side, one can steer into
(or under) the fall with the rear wheel, but then
the inertia (the so-called "centrifugal force")
will act to increase the rate of fall.

"Backward" configuration

The theoretical advantages cited above for rear-
wheel-steering bicycles can be fully realized
with a configuration in which the rider sits
backwards in relation to the direction of travel.
This arrangement permits the use of full-size
wheels and a short wheelbase (desirable for easy

Figure 9.10
Stability test of reverse-
direction bicycling.
Courtesy of Milton W.
Raymond.

balancing), yet does not require the rider to
straddle the steered leading wheel.

A low-speed-stability test machine was built
by M. W. Raymond (personal communication to
D.G.W.) to investigate the possibilities for a
roadworthy enclosed streamlined recumbent
"backward" bicycle (figure 9.10). The rider uses
a plane mirror for forward view. Although the
very low center of mass imposed the most severe balancing conditions, Raymond could balance at 0.5 mph (0.25 m/sec), and could perform
low-speed circles and figure-eights. Despite this
promise, he found that driving with a mirror
"far too repellent for normal use." When he
used a stabilized mirror to remove the dizzying
view caused by side-to-side weave, the rear-fac-
ing-rider concept still failed. At workable eye-
to-mirror distances, the widest mirror practical
for a streamlined housing gave too narrow a
field of view for true roadworthiness.

Another approach would be to have a forward-
facing rider steer both wheels, but the short-
wheelbase solo bicycle would be complicated
because the rider would straddle the front

wheel. The front wheel would have to be limited to making balancing corrections, while the rear wheel would do the gross steering, mainly in low-speed maneuvers. At the time of this writing (1981) Raymond was experimenting with a less complex method to steer both wheels of a long-wheelbase bicycle. For a first test of the wheelbase-immune balancing concept, Raymond and his co-workers have under construction a two-rider bicycle with a seat height of about 12 inches (300 mm) and a wheelbase of 12 feet (3 m). The relatively long streamlined hull required may bring crosswind effects that limit use of the concept to multi-rider bicycles.

Hands-off bicycling

Although riding "hands-off" is frowned upon by safety authorities, for obvious reasons, manufacturers of regular "safety" bicycles believe that their riding characteristics should be such that they can be ridden without the hands on the handlebars. One reason for not considering recumbent bicycles was the supposition that they cannot be ridden hands-off.[12] (In fact they can: Tom Winter, professor of classics at the University of Nebraska at Lincoln, who has constructed at least two recumbents, enjoys riding with arms folded when the situation is suitable).

Bourlet believed that there should be only a small sideways movement of the head part of the frame when the front wheel is turned to steer.[13] F.R.W. has speculated that the hands-off rider accomplishes the steering action by twisting the frame through this sideways degree of freedom, and that it is desirable that only a small amount of movement be permitted. Bourlet recommended a sideways movement as small as 20 mm, and gave a somewhat complicated formula relating the fork offset to the steering-head angle (the angle of the fork-rotation axis with the horizontal) for satisfactory steering. For a 27-inch wheel the calculated offset is 1.7 inches (45 mm) for a steering angle of

75°, giving a y/d value of 0.063. If such a calculation is made using Davison's simpler formula[14] for the same steering angle, the offset is again about 1.7 in. (45 mm). It will be seen from tables 9.1 and 9.2 that this confirms the approach of using the stability index of Jones (ref. 7), with the Bourlet and Davison formulas giving slightly lower stability indices than is usual nowadays. Davison felt it important that there be no rise or fall of the frame head for small steering movements, but because the wheel sweeps out (in general) a doughnut shape as the handlebars are turned, with the front wheel at the lowest point of the inclined doughnut at the neutral (straight-ahead) position, there will be no vertical head movement for small steering changes for any practical steering geometry in the range covered by tables 9.1 and 9.2.

References

1. S. Timoshenko and D. H. Young, *Advanced Dynamics* (New York: McGraw-Hill, 1948), p. 239.

2. J. P. Den Hartog, *Mechanics* (New York: Dover, 1961), p. 328.

3. G. S. Bower, Steering and stability of single-track vehicles, *The Automobile Engineer* V (1915): 280–282.

4. R. H. Pearsall, The stability of the bicycle, *Proceeding of the Institute of Automobile Engineering* XVII (1922): 395.

5. R. A. Wilson-Jones, Steering and stability of single-track vehicles, *Proceeding of the Institute of Mechanical Engineers* 17 (1922), no. 395: 191–213.

6. R. S. Rice and R. D. Roland, Jr., An Evaluation of the Performance and Handling Qualities of Bicycles, report VJ-2888-K, Cornell Aeronautical Laboratory, 1970.

7. D. E. H. Jones, The stability of the bicycle, *Physics Today* (April 1970): 34–40.

8. D. Banten and C. Miller, The geometry of handling, *Bicycling* (Emmaus, Pa.) (July 1980): 97–106.

9. J. P. Den Hartog, *Mechanical Vibrations* (New York: McGraw-Hill, 1956), pp. 329–334.

10. R. D. Roland, Jr., Computer Simulation of Bicycle Dynamics, paper, American Society of Mechanical Engineers, 1973.

11. L. H. Laiterman, Theory and Applications of Rear-Wheel Steering in the Design of Man-Powered Land Vehicles, B.S. thesis, Massachusetts Institute of Technology, 1977.

12. K. Hutcheon (Technical Director at T.I. Raleigh Ltd.), personal communication to D.G.W.

13. C. Bourlet, Le nouveau traité des bicycles et bicyclettes équilibre et direction, in *Encyclopédie scientifique des aide mémoire*, ed. M. Léaute (Paris: Gauthier-Villars, 1898), pp. 84–104.

14. A. C. Davison, Upright frames and steering, *Cycling* (3 July 1935): 16–20.

Recommended reading

E. Bernadet, L'Etude de la direction, *Le Cycliste* (September–October 1962): 228.

C. Bourlet, *La bicyclette, sa construction et sa forme* (Paris: Gauthier-Villars, 1899), p. 60.

R. N. Collins, A Mathematical Analysis of the Stability of Two-Wheeled Vehicles, Ph.D. thesis, University of Wisconsin, 1963.

CTC Gazette (February 1899), p. 73.

"Cyclotechnie," L'etude de la direction, *Le cycliste* (November 1972–January 1973).

F. DeLong, Bicycle stability, *Bicycling* (May 1972): 12–13, 45.

E. Dohring, Stability of single-track vehicles, Forschung Ing. Wes. 21 (1955), no. 2: 50–62 (tr. Cornell Aeronautical Laboratory, 1957).

———, Steering wobble in single-track vehicles, *Automobil technische Zeitschrift* 58 (1962), no. 10: 282–286 (MIRA translation 62167).

H. H. Griffin, *Cycles and Cycling* (London: Bell, 1890).

H. Fu, Fundamental characteristics of single-track vehicles in steady turning, *Bulletin of Japanese Society of Mechanical Engineers* 9 (1965), no. 34: 284–293.

M. Kondo, Dynamics of Single-Track Vehicles, report, Foundation of Bicycle Technology, 1962.

G. T. McGraw, *Engineer* (London) 30 (2 December 1898).

J. R. Manning, The Dynamical Stability of Bicycles, report RN/1605/JRM, Department of Scientific and Industrial Research, Road Research Laboratory, Crowthorne, England, 1951.

R. S. Sharp, The stability and control of motorcycles, *Journal of Mechanical Engineering Science* 13 (1971), no. 4.

D. V. Singh, Advanced Concepts of the Stability of Two-Wheeled Vehicles: Application of Mathematical Analysis to Actual Vehicles, Ph.D. thesis, University of Wisconsin, 1964.

A. Van Lunteran and H. G. Stassen, Investigations of the Characteristics of a Human Operator Stabilizing a Bicycle Model, International Symposium on Ergonomics in Machine Design, Prague, 1967, p. 27.

———, On the Variance of the Bicycle Rider's Behavior, Sixth Annual Conference on Manual Control, Wright-Patterson (Ohio) Air Force Base, 1970.

D. M. Weir, Motorcycle Handling Dynamics and Rider Control and the Effect of Design Configuration on Response and Performance, University of California, Los Angeles, 1972.

F. J. W. Whipple, The stability of the motion of a bicycle, *Quarterly Journal of Pure and Applied Mathematics* 30 (1899): 312–348.

Materials and stress

The makers of early bicycles used "traditional" materials: woods reinforced with metals, as used in the earliest vehicles. The shortcomings of this type of construction for human-powered vehicles soon became apparent, and tubular-steel construction with rolling (instead of rubbing) bearings appeared in the 1870s. In general there has been no basic change in these basic principles of bicycle construction, although smoother roads, better steels, aluminum alloys, and improved design have resulted in a reduction in bicycle weights to about one-third of that common for early machines.

For most of the past century, then, the principal materials used for the frames of bicycles have been steels: low-carbon for inexpensive machines, medium-carbon for the middle-range models, and heat-treated chrome-molybdenum-manganese-carbon alloys for the best competition cycles. Inexpensive frames are made of straight-gauge tubes formed from steel strip, rolled and electrically welded along the seams, and later welded to the other frame components. The best frames are made from seamless tubes, drawn to be thinner in the middle than at the ends, and silver-brazed into close-fitting tapered end sockets called lugs.

As a result of commercial development in general, and of aerospace activities in particular, many new materials and combinations of materials are available. Some of these materials are being tried out for bicycle frames and components. Some show advantages over traditional steels, yet the lack of a clear understanding of the requirements of these materials may lead to disappointment and even danger. This chapter will draw attention to the most important properties to be looked for in new materials and give some guidance in their use for bicycle frames and for other components.

Factor of safety

All structures are designed to be stronger than is strictly necessary in normal service. The ratio between the load that would cause a structure to fail and the normal load is called the *factor of safety*. Bicycles are built with large factors of safety. They are often used to carry grossly oversized loads. Fairly standard bicycles are used in circus acts to carry five or ten people or even more. A well-known advertisement showed fifteen men carried by a commercially available bicycle. Yet bicycles fail. Spoke breakage is perhaps the most common failure. Aluminum-alloy handlebars and stems all too frequently break off. In some makes and models of bicycles there have been a series of failures of the front forks, with often unfortunate consequences.

Loading

How can these failures be reconciled with the large factors of safety? There is ignorance about two vital matters: how a bicycle will be loaded throughout its life and how the pattern of loading may contribute to eventual failure.

The bicycle designer cannot assume that all users will treat their machines with care and attention at all times. Some will ride them up and down curbs, possibly with friends sitting on the handlebars, the crossbars, or the carrier. Some users will be unable to avoid deep potholes in the road. Some will bolt heavy toolboxes to the frame or to the carrier. The degree to which these practices constitute use or abuse has become generally recognized as being different for different types of bicycle. At one extreme is the American "cruiser," with its heavy construction, large balloon tires, and single low gear, which is designed to be ridden at speed over curbs and other hazards. At the other extreme is the lightweight racing or track bicycle, which will crumple under extraordinary stress and which runs on tubular tires (perhaps with silk reinforcement) intended to last just as long as the race.

Under all these different conditions, the loading which most of the components of a bicycle experience can be very complex. The tubes in a frame are usually loaded under a combination of bending, shear, torsion, and tension or compression. Appropriate sizes for these components have been arrived at by experience, not by analysis and prediction. Even with advanced computers it would be difficult and expensive to analyze all the combined stresses in a bicycle frame and thereby to improve its design more than marginally. Given this situation, the best approach to the use of new materials and new different configurations is to compare them with the strength and other characteristics of present successful designs.

Strength of materials The word "strength" must be interpreted very carefully. One reason why a large factor of safety implies ignorance is that the wrong criterion of failure is often used. For instance, the factor of safety is often based on the maximum strength of the material—the "ultimate tensile stress" (UTS). Yet most failures occur not because the material has been exposed to a stress greater than the UTS (usually it can be demonstrated that the stress at failure is only a fraction of the UTS), but because of "fatigue."

Fatigue

Metals, and other materials, "tire." We all know that we can break a soft metal wire by bending it so that it takes a "set" and then bending it back several times. A failure brought about by this type of loading is called *low-cycle fatigue*. Materials also fail by being stressed many more times through much lower stresses, in so-called *high-cycle fatigue*. By subjecting tens or hundreds of specimens of different materials to cyclic stresses on special fatigue-testing machines, starting at a high stress and gradually lowering it, curves similar to those in figure 10.1 are obtained.

A "touring" wheel can be expected to last 20,000 miles, about 15 million rotations. The spokes should be in tension continuously, but the tension will vary between a maximum and a minimum once per revolution. Mny spokes will fail through fatigue well before experiencing 15 million stress cycles, but spoke failures are in the range of high-cycle fatigue. Most failures of other bicycle components (such as brake cables, frames, and handlebars) are midway between low-cycle and high-cycle fatigue, with stresses applied tens of thousands rather than tens of millions of times.

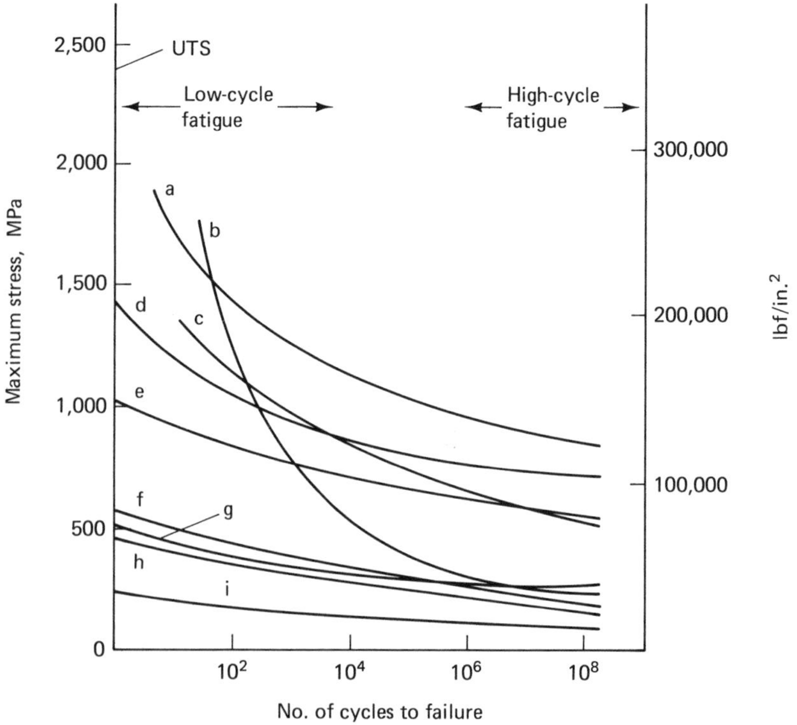

Figure 10.1
Endurance limits of various materials. (a–c) Composites: (a) Kevlar 49 and epoxy, boron and epoxy; (b) "S" glass and epoxy; (c) Graphite and epoxy. (d) 4130 Cr-Mo alloy steel. (e) Titanium alloy IMI318. (f) Aluminum alloy 7075T6. (g) Medium carbon steel. (h) Aluminum alloy 2024T6. (i) Magnesium. Curves d and g are endurance limits for steels.

No. of cycles to failure

There is evidence in figure 10.1 of two crucial differences in the fatigue behaviors of different materials.

The fatigue curves for steels show falling permissible stresses as the number of loading cycles increases, but eventually the curves appear to flatten out. When they do, the stress reached is called the "fatigue limit" or the "endurance limit." This means that if the designer uses a factor of safety to ensure that the stresses in a component never rise above that limit, its life will be theoretically infinite no matter how many stress reversals it is subjected to. Nonferrous metals like aluminum, titanium, and magnesium alloys and composite materials do not show an endurance limit. The more stress reversals are imposed, the lower is the permissible stress. (Some experts state that this is also the case for steels, but that the fatigue-stress curves have a very small slope.) Tables of material properties, for instance table 10.1, rather misleadingly show a fatigue-limit stress for nonferrous materials. In fact, this is usually the stress that will allow 500 million stress reversals to be withstood.

The second important difference is that the fatigue-limit stress is about half the ultimate tensile stress for steels, but may be as low as 16 percent of the UTS for some composites (table 10.1). The UTS is, therefore, a poor guide to material suitability. If the application will involve far fewer loadings than the fatigue-limit stress no matter how long the component is used, then the use of this stress (with a factor of safety) is acceptable. But there are many applications where 500 million reversals, large though this number seems, can be easily and quickly exceeded—for instance, when a component is vibrating at a fairly high frequency. Something vibrating at 3 kHz (3,000 cyclic reversals per second) will experience 500 million reversals in 50 hours.

Table 10.1 Properties of bicycle materials, compiled from various sources.

	Modulus of elasticity, E (GPa)	Ultimate tensile strength (MPa)	0.2% proof stress at yield (MPa)	Elongation at failure (%)	Fatigue limit/UTS (5×10^8 cycles)	Density (Mg/m$_3$) (specific gravity)
Steels						
Medium-carbon	200	520	310	26	0.5	7.85
CrMo (AISI 4130)	200	1,425	1,240	12	0.5	7.85
Aluminum alloys						
2024-T4	73.1	470	325	20	0.29	2.80
6061-T6	68.9	310	276	12	0.31	2.80
7075-T6	71.7	570	503	11	0.265	2.80
Magnesium	44	248	200	5–8	0.37	1.79
Titanium alloys						
IMI 125 (pure)	105–120	390–540	340	20–29	0.5	4.51
IMI 318	105–120	~1,000	900	8	0.55	4.42
Composites[a]						
"S" glass-epoxy	90	3,750?	3,450	3.5?	0.16	2.63
HT graphite-epoxy	221	3,600?	2,000	1.25	0.25?	1.75
Boron-epoxy	250?[b]	1,200	?	?	0.80?	1.90
Boron-aluminum	165	1,025	?	0.65	0.70	2.40
Kevlar-49-resin	75	1,380	?	2.75	0.70	1.45
Glass-nylon (Dupont Zyftel FE 8018 NC-10)	2.3	59.9	59.9	14	?	1.18
Woods (ash, beech, oak)	12	100	60	?	?	0.67

a. Composite properties are lower (sometimes much lower) in compression and in tension in directions across the fiber alignment. Properties vary greatly depending on the materials used and on the mode of failure.
b. Question mark indicates uncertain or scattered data.

Figure 10.2
Fatigue testing of bicycle frame. Rollers incorporate bumps; masses simulate rider loading. Rotating masses on cranks add cyclic twisting moment on frame, approximating pedaling torques. Courtesy of Raleigh Industries, Inc.

The fact that handlebars, for instance, break fairly often (D.G.W. has had three fail on conventional sports bicycles), and that they do so at far fewer than 500 million reversals, shows that they have been poorly designed and that the builder probably had in mind the UTS rather than the fatigue stress. We conclude that, because no important component of modern bicycles vibrates at high frequency, the use of the published fatigue limits is acceptable for nonferrous as well as for ferrous materials, and for nonmetallic materials. We also emphasize that the fatigue-limit stress rather than the ultimate tensile stress should be used as a criterion of acceptable strength.

Samples of mass-produced bicycle frames, and frames using new materials or construction methods, are subjected to fatigue tests such as that illustrated in figure 10.2. The heavy-duty wheels (which are not being evaluated) run on rollers incorporating bumps, which can, if desired, be made to simulate cobbled streets. Masses on the seat post, on the crank ends, and on the handlebar stem simulate rider loads. A fixed gear is used so that the spinning cranks impose a fluctuating twisting load on the frame. The test shown in figure 10.2 is being performed on an experimental frame in which the three main tubes are graphite composites with compression-clamped lugs. Builders of special high-quality frames cannot undertake the expensive testing illustrated here, and it is important, therefore, that they adhere closely to well-established practice.

Increases in stress

Another reason why some "alloy" handlebars break is that the manufacturers have not taken into account the serious effect of sudden changes of section on the local peak stress. At the point where the handlebar is clamped into the stem, the bending moment from the forces exerted by the rider's arms at the bar ends will be at a maximum. If the clamp on the handlebar stem fits well, it will act as if the handlebar had a sharp change in cross-section. This can multiply the already high stresses manyfold, by what is known as a *stress-concentration factor*. Figure 10.3 shows the stress-concentration factor for bending of a simple rod which has a sharp change in diameter (taken from reference 1). The maximum stresses are affected more by the radius of curvature at the junction than by the diameter ratio of the section change.

Changes of section should, therefore, be gradual. High-quality frame lugs are tapered and filed with decorative patterns to transfer the stress gradually from tube to lug and then to a connecting tube. Likewise, high-quality frame tubes are gradually tapered in wall thickness to provide thicker walls at the ends, where the bending stresses are highest. Application of the same principles to handlebars would greatly decrease the danger of breakage. Two alternative

Figure 10.3
Stress-concentration
factors.

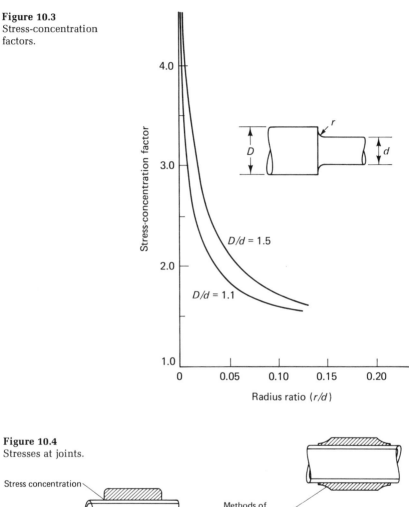

Figure 10.4
Stresses at joints.

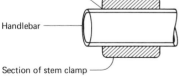

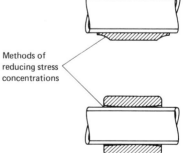

methods of greatly reducing a serious stress concentration at the handlebar stem clamp are shown in figure 10.4.

Other material properties and criteria for choice

Being strong enough is a necessary but not a sufficient condition of suitability of a material for bicycle construction. Some other requirements are these:

• The density of the material must be such that the resulting structure is light.

• The resulting structure should not be unduly flexible. (The property defining flexibility in table 10.1 is the elastic or Young's modulus, E.)

• The failure should be gradual rather than sudden. (One property that gives some indication of the failure mode is the elongation at failure, in table 10.1.)

• The cost must be reasonable.

• Joining one piece to another should be possible without loss of strength in the parent material(s) or in the joint.

• The material should intrinsically resist, or should be easily protected from, corrosion.

The limiting fatigue strength, the elastic modulus, and the density are probably the most important physical properties for bicycle materials. Only a little less significant is the elongation at failure. This is the permanent "set" a specially machined circular rod will take when it is pulled in a tensile-testing machine until the rod breaks. A brittle material like glass has no elongation at failure; a soft material like lead will stretch considerably before failing.

Yielding

Low- and medium-carbon steels (up to about 0.2% carbon) exhibit the special characteristic of *yielding*—stretching and then going on to take somewhat higher stress before ultimate failure

Figure 10.5
Yielding of steels.

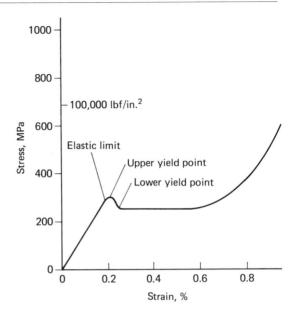

(figure 10.5). Yielding allows steel structures to accommodate local high stresses, thus lowering the stress locally and allowing other parts of the structure to take up more of the load. This "forgiving" nature of steel has permitted thousands of structures to be classified as successful whereas, if they had been made in a material not exhibiting yield, failures would have propagated from regions of high stress to the remaining components, which would have started at low stress levels but then would have exceeded their UTS.

The yielding of a component is often visible, especially in bending such as that of an overstressed front fork. Obvious deformation or the cracking or flaking of enamel warn of high stresses and of possible incipient failure.

High-strength alloy steels, such as the new Reynolds 753, do not have defined yield points. Therefore, rear forks made of these alloys cannot be given a "set" (a permanent deformation) to accommodate additional cogs, for instance.

The presence or absence of yielding in a material does not necessarily affect the nature of fatigue failures, which almost always occur without warning. (In components with very low mean stresses, like mudguards or fenders, fatigue cracks may grow slowly through the material, giving ample warning. Unfortunately, this benign process cannot occur in components where the mere initiation of a fatigue crack will increase local and mean stresses to a level at which the crack will almost instantly spread right across the section.) A series of disastrous failures in the front forks above the fork crown of a certain make and type of bicycle were traced to incorrect heat treatment coupled with a sharper-than-specified radius of curvature, which raised local stresses.

Most fiber-composite materials fail suddenly, whether at their ultimate stress or through fatigue failure. As stated in reference 2, "The survived specimens do not show any strength degradation. This . . . indicates that the fatigue failure is like a sudden death; that is, the fatigue failure occurs without any visible evidence of damage."

Notch sensitivity

Hard or strong materials are often notch-sensitive in the way that glass is sensitive to a scratch. Glass is cut to size by making a small scratch and applying a very small fraction of the normal breaking load. Soft materials can also be notch-sensitive. It is common experience that some plastic films and some synthetic-rubber inner tubes will tear easily once a small cut has been made, while other films of approximately equal strength and thickness will "heal" a cut and greatly resist tearing. Notch-sensitive materials are not suitable for use in bicycle components, which are likely to be scratched in normal use.

Notch sensitivity can be measured in testing machines in which a notched specimen is struck by a swinging weight, the energy loss of which is recorded. In metals, notch sensitivity is a function not only of the material but also of the heat treatment. A material that is not notch-sensitive is called "tough." Steels, aluminum alloys, and fiber composities are normally very tough. Should the use of a new material be contemplated for use in a bicycle frame or components, its fracture toughness or notch sensitivity should be compared with that of steel before development work is undertaken.

Joining properties
The ends of a frame member, where the bending moments are normally highest, are also the points at which the member must be joined to other members. The means for joining one member to another must, therefore, be sufficient to take the loads involved and must not degrade the material properties. High-quality lightweight steel frames use lightweight tubular angled sockets, called "lugs," into which the frame tubes are brazed. Low-alloy-steel frames can be brazed with regular brass or bronze brazing alloy, but some high-strength materials require the use of silver-alloy brazing solders so that the heat-treatment temperature limit is not reached.

Welding, used on the cheapest steel frames and on most aluminum-alloy frames, melts a small quantity of the material into the joint, either from a separate weld-metal rod or from the parent metal. All heat treatment in the weld region is lost, and in addition the alloy constituents will be somewhat uncertain. The shrinkage of the solidifying and cooling metal will introduce thermal stresses. Aluminum frames must therefore be heat treated, first to relieve these stresses and then to restore most of the original properties to the metal. Thick-walled tubes are used in welded steel frames so that stresses are low and heat treatment is not necessary. The ability of steel to yield gives an additional safety factor.

Aircraft wings and parts of fuselages have been glued together with high-strength adhesives in highly controlled circumstances. With suitable close-fitting lugs this procedure should also be satisfactory for aluminum bicycle frames, with no degradation of properties, no thermal stresses, and considerable savings of time, costs, and energy. Some high-quality Italian frames are made in this way; the tubes and the lugs are threaded for added reliability.

A combination of adhesives and lugs, sometimes involving compression, has also been used for the fiber-composite frames tried out recently (for example, the experimental frame shown in figure 10.2).

The frame of the human-powered aircraft *Gossamer Albatross* was constructed of carbon-fiber composite tubing. The joints were made by wrapping adjacent butting tubes with "prepreg" (resin-impregnated) tape and then heat treating.

Recently there has been a revival of a method of producing steel-tube bicycle frames developed by an early British manufacturer in 1902: the casting of aluminum lugs around jig-held joints.

Corrosion resistance

Nonferrous metals and plastics are more resistant than steel to atmospheric corrosion. The surface treatments necessary to ensure satisfactory service are minor operations compared with the plating or enameling processes required for steels. On this account, the use of the newer materials for the less stressed parts of bicycles has been generally satisfactory and will no doubt be extended in various ways. It is interesting to note that celluloid mudguards were in use in Victorian times, and aluminum components also appeared and disappeared.

Cost

Cost was ignored in the above but it usually affects the choice of materials. At present, steel is the least expensive material for making a bicycle. When the cost of manufacture is considered, it is possible that high-strength plastics may win a place because of the automated production these materials allow.

Steel versus aluminum alloys

From the foregoing discussion of material properties it can be appreciated that the design of the major components of a bicycle, starting from the known (desired) configuration of the machine and a fair knowledge of the applied loads, would be complex. However, we are fortunate to have examples of successful components such as steel-tube diamond-pattern frames before us, and we can simply compare with them the size and mass of the same components produced in alternative materials to give the same performance.

Let us suppose that we wish to compare the weight of an aluminum-alloy frame designed to have the same strength and stiffness as a steel frame. Although we know that the loading of a bicycle frame can be complex, we choose simple bending as the loading method because it will serve well when we are just comparing one material with another. Both frames will be constructed from circular tubes. We will ignore for the moment the question of joining the tubes.

Any standard engineering reference book, such as reference 3 or 4, will give the stiffness (force per unit deflection) of a cantilevered beam as

$$\frac{3EI}{L^3},$$

where E is the modulus of elasticity, I is the section moment of inertia, and L is the length of the beam. For a circular tube,

$$I = \frac{\pi D^4}{64} [1 - (d/D)^4],$$

where D is the outside diameter of the tubing and d is the inside diameter.

For steel and aluminum tubes of the same length to have the same stiffness, the product ED^4 must be the same for both materials if the diameter ratio d/D is, at least for the moment, specified as identical for the two materials. Table 10.1 shows that the modulus of elasticity of aluminum alloy is about one-third that of steel. Thus, the tube diameter of the aluminum-alloy frame must be $3^{1/4} = 1.316$ times that of the steel frame.

Now we must ask this question: If the frame is as stiff as the steel frame, will it be safe from fatigue failure? The maximum stress in a circular tube for a specified load and tube length is given by the relation

$$\frac{\text{Max. stress}}{\text{Load}} = \frac{32L}{\pi D^3[1 - (d/D)^4]}.$$

Therefore, the maximum stress in an aluminum frame of equal stiffness to a steel frame is $1.316^{-3} = 0.439$ of the peak stress in a steel frame (again, for the same ratio of inside to outside diameter).

The fatigue-limit stress in the strongest of the three aluminum alloys listed in table 10.1 is

$0.265 \times 570 = 151$ MPa.

The fatigue-limit stress in the steel-alloy frame is $0.5 \times 1{,}425 = 712.5$ MPa. Therefore, the ratio of the fatigue-limit stresses is 0.212, which is much less than the ratio of peak stresses (0.439), and the aluminum-alloy frame will be much more highly stressed (perhaps dangerously so).

The weight of the two frames would be proportional to ρD^2, where ρ is the density:

$$\frac{\text{Weight of alum.-alloy frame}}{\text{Weight of steel-alloy frame}}$$
$$= \frac{(\rho D^2)_{\text{al.}}}{(\rho D^2)_{\text{steel}}} = \frac{2.80}{7.85}\sqrt{3} = 0.618$$

An alternative criterion is to design the aluminum-alloy frame to have peak stresses that are the same proportion of the fatigue-limit stresses as for the steel-alloy frame. Then,

$$\frac{D_{\text{alum.}}}{D_{\text{steel}}} = \left(\frac{712.5}{151}\right)^{1/3} = 1.677.$$

The ratio of the weights of frames made from the two materials would be

$$\frac{\text{Weight of alum.-alloy frame}}{\text{Weight of steel-alloy frame}}$$

$$= \frac{2.80}{7.85}(1.677)^2 = 1.003.$$

Therefore, fortuitously, the weights have turned out to be virtually identical in the two frames when designed for the same proportional fatigue-limit stresses. The aluminum-alloy frame would, however, be much stiffer. (An aluminum-alloy-frame road-racing bicycle is shown in figure 10.6). The designer therefore has some freedom to trade off among stiffness, stress, and

Figure 10.6
Klein road-racing bicycle with aluminum-alloy frame. Courtesy of Gary Klein.

weight by changing not only the tube diameter but the ratio of inside to outside diameter. A track bicycle, to be exposed to few bumps and potholes, could well be designed to an equal-stiffness criterion in aluminum rather than steel. It would seem unsafe to do so for touring bicycles, which often are loaded with heavy bags and travel on rough streets—ideal conditions for building up fatigue damage.

This illustration was meant to be just an example of how to use material-property data in design. The important principles are to use fatigue-stress limits rather than ultimate tensile stress, to consider stiffness as well as strength, and to take successful components as models of stiffness and strength because of the great uncertainty in the actual magnitude, type, and frequency of loads.

Nonmetallic components

Fabricators in plastics, particularly, have lately made great advances toward producing machine parts competitive with metal parts where quiet running, low price, and light weight are important factors. If corrosion resistance matters greatly, as in chemical plants, nonmetallic parts may have considerable advantages over metal ones.

For most conditions of bicycle use, some plastic components show serious drawbacks when compared with corresponding metal parts. Plastic bearings must be made with larger clearances than plain metal bearings; that is, the fit is "sloppier." Nylon 66, a hard plastic, is one of the most slippery materials from which to make a bearing, but its minimum coefficient of friction of 0.04 shows a fourfold greater resistance to movement compared with a reasonably good ball bearing's performance of 0.01. Several firms now make reinforced-plastic chains and reinforced-rubber toothed belts. All need to be larger, and some heavier, than a modern steel chain. The chainwheels would be very wide (15-mm or ¾-inch teeth at least) and hence cum-

Figure 10.7
Bicycle wheels of glass-reinforced nylon ("Zytel"). Courtesy of E. I. du Pont de Nemours and Co.

bersome. Nylon derivatives are successfully used for motorcycle rear-wheel sprockets, and could perhaps be used for bicycle sprockets. However, it appears, because of the low strength of the material compared with steel, that a nylon hub gear would be much bulkier than the standard steel hub gear.

Glass-reinforced nylon is being used commercially for youth off-road bikes with apparent success (figure 10.7). The higher weight and lower stiffness would not make such wheels attractive for racing. An experimental wheel was made at MIT by Kindler, who used a filament-winding technique to make the hub, spokes, and rim entirely from resin-soaked "Kevlar" (figure 10.8).[5] This high-modulus DuPont fiber (polyphenylene terephthalamide) has a strength greater than that of steel, and has to be regarded as promising for future bicycle construction.

In 1955 the well-known cycle designer Cohen expressed great enthusiasm for PTFE bearings.[6]

Figure 10.8
Kindler's Kevlar wheel.
From reference 5.

However, it has been found that the compressibility of that plastic causes a great deal of trouble. Plastics of various types have since been used for bearings in children's cycles, and some complete parts (such as small pedal frames) of plastic have been marketed. Children do not seem very concerned with easy running in their bicycles and tricycles, and parents are not unhappy if their children are slowed down. However, manufacturers appear to have realized that adult cyclists will not accept plain bearings of plastic (or the plain metal bearings recently tried in pedals). There seems to be a consensus that ball bearings are essential in an adult's machine to ensure easy running and reasonably long life without constant adjustment. (There is one exception: The lightly loaded "jockey pulleys" in most derailleur gears have been produced with plastic bearings by most manufacturers). No doubt the cyclists of the 1870s through the 1890s were glad to see plain bearings go, and present-day veteran-cycle enthusiasts will endorse that opinion.

Alternative frame materials

Woods

Bicycles with wooden frames have been made and ridden with satisfaction at regular intervals since the earliest "hobby-horse" days around 1800. The Macmillan rear-drive bicycle introduced in 1839 was followed by a large number of "boneshaker" front-drivers from about 1860. In the 1870s metal construction became dominant, but there were regular resurgences of wood frames (including some of bamboo) until the end of the century (see figure 10.9). The Stanley shows of this period included bicycles with completely wooden wheels fitted with pneumatic tires; an early Columbia with such wheels is on display at the Science Museum in London. Various wooden-framed bicycles dating back to the 1890s are still ridden by proud owners in veteran-cycle rallies.

Although wood was used regularly up until the 1930s for rims (for both sew-up and clincher tires), and wooden mudguards and seat pillars were not unknown, the wooden frame did not appear again until the 1940s, when metal was needed for the war. However, wood became scarcer than steel. An American wooden-framed

Figure 10.9
Bamboo-frame bicycle.
From reference 7.

bicycle from the period is on show in the Washington Museum.[7] A cane-framed bicycle appeared in Trieste in 1945; Wilde[8] thought it to be a sound proposition and stated that it was rigid enough to ride up hills.

Molded plastics

Since the recent advent of relatively large moldings in plastics (sometimes reinforced), there have been several attempts (at least one of them fraudulent) to market molded bicycle frames. As figures 10.10–10.12 show, these frames can be rather bulky-looking (more so than the bamboo and other wooden-framed bicycles marketed over the years). The lack of popularity of these molded frames is due partly to this. Most were also very flexible in comparison with steel-framed bicycles. As new polymers and polymer-fiber combinations are developed, plastic frames will become less bulky, lighter, and stiffer. There are certainly advantages for general everyday use in a frame made of an inexpensive material that is completely resistant to corrosion.

Plastic tubes

The bulky shape of the molded plastic frame can be avoided if the frame is constructed along conventional lines, using tubes fitted into joints. Only recently have nonmetallic materials been made that, in tube form, could approach metals if weight and bulk were taken into account. Such materials are plastics reinforced with carbon or Kevlar fibers, which are now sold at reasonable prices and have higher tensile strengths than strong steels and high Young's (elasticity) modulus values. The fibers do not, however, exhibit one of the desirable properties of metals: They do not stretch appreciably before breaking. Also, the composite fiber structures have different properties "across the grain" than "with the grain," and the fibers have to be embedded in a resin that is very weak by comparison, giving a

Figure 10.10
American fiberglass-
frame bicycle, 1963.

Figure 10.11
British plastic-frame
bicycle.

Figure 10.12
Dutch plastic-frame
bicycle. Courtesy of
Plastics and Rubber
Research Institute.

composite of varying properties, most much less
attractive than those of the fiber. Although the
properties of carbon and Kevlar fibers are well
known, the properties of usable forms such as
tubes are not. In particular, the fatigue strength
of composite-fiber tubes has not been well docu-
mented. In general, fiber-composite materials
have very poor compressive strengths (often
one-tenth of the UTS) and poor shear strengths.
However, according to one major manufacturer,
a Grafil composite tube weighs less than a light-
alloy tube of similar strength. The most unattractive feature of these tubes is that the joints
have to be in the form of clamps (see figure
10.2). Adhesive joints are considered too weak,
and any drilling and riveting is liable to cause
failure without warning.

Aluminum
The first innovations in nonferrous metals for
frames occurred in the 1890s. Humber used aluminum tubing with lugs, and a French manufacturer introduced a "Lu-mi-num" bicycle of
cast alloy. The Beeston Humber frame was reported as very satisfactory, and it was said that
the whole machine—with gear case, lamp, and
tools—weighed only 27 lb (12.3 kg).[9] There are
no easily available records about the Lu-mi-num
machine, but a table of tests published in the
journal *Engineering* gave a strength comparison
with a current steel frame.[10] This table shows
that the balance for ultimate strength was in favor of the steel frame. The fatigue strength of
the aluminum frame would presumably have
been even more unfavorable.

Many other types of aluminum frames have
appeared on the market from Continental factories. Because aluminum brazing was not practicable, various designs of lugs were used to
grip the tubes via corrugations or internal plugs.
(A design using threaded and glued lugs was
mentioned above.) The most recent clamping
type of lug can be seen on the Caminargent of

the 1930s, which used octagonal tubing. In addition, various welded-joint frames have been marketed, in spite of the effects mentioned above. The better examples have, of course, been heat-treated after welding, such as the high-quality Klein frame of figure 10.6.

Nickel

Nickel tubing followed the use of aluminum in the 1890s, no doubt in an attempt to produce a rustless frame. The firm manufacturing the frames, however, existed for only a short while during the bicycle-boom period when cost was of less importance. Nickel was and is more expensive than steel, but it is strong and rigid and can be welded satisfactorily. It is seldom used in its pure form but is a major component, with chromium, of stainless and high-strength steels.

Titanium

The history of aluminum bicycle frames is being repeated in the case of the recent use of titanium. Less than a decade since the first commercial production of a once very costly metal, it is being thought of seriously for bicycle frames. Titanium in various alloy compositions is now used for corrosion-resistant heavy engineering equipment, for the spars and skin of high-speed aircraft, and for the disks and blades of jet-engine compressors. Satisfactory welding methods using inert-gas shielding to avoid weld deterioration have been developed. For bicycle use, titanium is corrosionproof.

Because titanium has a specific gravity about half that of steel, it was possible for the firm of Phillips to make of it a fairly conventionally shaped bicycle frame weighing 2¾ lb (1.25 kg), displayed at the London Cycle Show in 1956. No models were offered for sale, but the price would have been high. In 1973 the Speedwell Gear Case Co. of Birmingham, England, produced 10,000 frames with titanium tubing and sold them for about £130 or the equivalent. The mass of the frame and fork was advertised as $3\frac{3}{4}$ lb (1.7 kg).

Magnesium
The only other metals likely to be considered for bicycle frames are magnesium and its alloys. They have an attractively low specific gravity of about 1.7, which to some extent compensates for the relatively low tensile strength, and for the very low modulus of elasticity, which is one-fifth of that of steel. An alloy termed Elecktron was used fairly satisfactorily for making bicycle rims in the 1930s, but there have been no further applications in cycle manufacture.

Conclusions and speculations

Although much experience has been accumulated in the manufacture of bicycle frames and accessories in steel and aluminum alloy, and in the production of low-stress parts such as mudguards (fenders) in relatively well-known nonmetallic materials, there are incentives to try out new substances. The desired properties are lightness, corrosion resistance, low price, and the possibility of making the frame in a single piece. There might be advantages in using unit-construction methods for metals which avoid machining, such as lost-wax precision casting.

We can expect improvements in frame design and manufacture to give greater torsional stiffness. Such improvements would make acceptable a one-piece standard-type frame of plastic reinforced with carbon fiber. This could avoid the use of bulky and weak joints and take full advantage of the strength of fibers.

Efficient designs for a lightweight bicycle wheel and frame were evolved by the 1870s. The tension-wire-spoked suspension wheel and the hollow-metal-membered brazed-joint frame had by then ousted all other designs. These designs were pioneered by the cycle industry and not copied from some other branch of engineering. This pioneering led to the establishment of specialized industries, such as steel-tube manu-

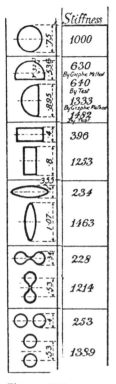

Figure 10.13
Relative stiffnesses of
equal-weight tube
shapes. From R. A.
Garrett, *The Modern
Safety Bicycle* (London:
Whittaker, 1899).

facture, and in addition accelerated progress in
the ball-bearing industry; so, in a significant
way, it helped to launch the aviation age.

Engineering science and practical considerations established the closed-section frame member, mostly of round or near-round section,
assembled with rigid joints generally incorporating lugs. (The stiffness of tubular sections
relative to round sections of the same mass per
unit length is shown in figure 10.13.) There was
rapid rejection of the practice—common in
other areas of structural engineering—of using
channel-section members bolted at the joints.
The optimum frame shape for a safety bicycle
also came early in the progress of the bicycle
industry in the form of the Humber pattern,
now called the diamond frame. Before this standardization came about in the 1890s there had
been a multitude of frame patterns, most constructed of much more robust and heavier tubing than that desired in the 1890s when only
reasonably light machines were acceptable.
Some of these early frame designs appear nowadays in children's machines or special machines
for adult use, such as small-wheelers. Oval
tubes (figure 10.14) are being reintroduced for
aerodynamic reasons.

Evidence that the diamond frame is of near-optimum shape and thickness is given by the fact
that track-racing and other bicycles can be built
that weigh from $6\frac{3}{4}$ to about 10 pounds (3–4.5
kg) (see figures 10.15 and 10.16 and table 10.2).
The steel-tubed frames of these machines are
very light indeed, showing that there is a good
approach to a minimum of metal and hence a
good placement of the members. It is interesting
to compare the above weights with the weight
of $8\frac{1}{2}$ pounds (3.85 kg) of a pair of pneumatic-
tired roller skates of the type shown in figure
10.17. These skates represent a high degree of
"weight paring" for a wheeled human-propelled
machine, and yet complete bicycles have been
made lighter.

Figure 10.14
Bottom bracket of
Raleigh lightweight
bicycle with oval tubing.
Courtesy of T. I. Raleigh,
Ltd.

Figure 10.15
Raleigh Professional
track cycle, 1974.
Courtesy of Raleigh
Industries of America,
Inc.

Figure 10.16
Tribune bicycle, 1895.
Perhaps the lightest
adult-size standard
bicycle ever built, this
machine weighed 8 lb 14
oz. From *Riding High:
The Story of the Bicycle*
(New York: Dutton,
1956).

Figure 10.17
Pneumatic-tired roller
skate.

Table 10.2 Lightweight bicycles with steel frames.

Date	Weight, lbm	Type	Make	Material of construction
1888[a]	$15\frac{1}{2}$	Cross-frame safety, solid rubber tires	Demon	Steel
1888[a]	19	Diamond-frame safety, solid rubber tires	Referee	Steel
1888[a]	11	High ordinary	James	Steel
1895[b]	$8\frac{7}{8}$	Diamond-frame safety, Pneumatic tires	Tribune	Steel
1948[c]	$8\frac{5}{8}$	Modern track bicycle	Legnano	Steel & Alloy
1949[d]	$6\frac{3}{4}$	Modern track bicycle	Rochet	Steel & Alloy

Sources:
a. *Bicycling News,* 8 February 1888.
b. *Riding high: The Story of the bicycle* (New York: E. P. Dutton & Company, 1956), p. 125.
c. *Cycling,* 7 January 1948, p. 10.
d. *Cycling,* 3 November 1949, p. 514.

Figure 10.18
The Dursley-Pedersen luxury bicycle, circa 1907.

Sharp gave many examples of the calculation of stresses in cycle frames for the more simple static loadings.[11] It appears that the use of standard modern lightweight strong steel tubing of near 22 gauge provides a safety factor above the yielding point of about 3 for distortion of the bracket through full-weight pedaling. The safety margin for simple vertical loading on the saddle pillar is very large.

Even after the diamond-framed safety bicycle had become accepted as standard, other types were introduced for special purposes. A radical example is the Dursley-Pedersen luxury bicycle (figure 10.18), whose frame design, patented in the U.K. in 1893, was triangulated so that tubular members were supposed to be subjected to compression stress only and other members (cables) withstood tension only. (Some tubes would, of course, have the twisting moment resulting from the rider's downward force on the pedals, the forces on the handlebars, and the road reaction through the wheels, and some tubes would be in tension.) Light versions weighed 14 lb (6.4 kg).[12] The hammock-style seat was an admired feature of this advanced bicycle.[13] Dursley-Pedersen frames are currently (1981) being made in Denmark.

An ingenious folding bicycle, also using tubes in compression and stainless-steel cables in tension, is the Hub and Axel "Pocket Bicycle" shown in figure 10.19. This is probably the most rigid available folding machine.

A very successful (but flexible) collapsible bicycle is the Bickerton shown in figure 10.20, which is predominantly constructed of aluminum alloys and has an open frame for ease in folding.

The modern trend in open frame construction was perhaps started by the Moulton. An attractive selling feature is that this type of bicycle can be used by either sex in any dress. Open frames must be made with a large main frame member if undue flexibility is to be avoided.

Figure 10.19
Foldable "pocket" bicycle. Courtesy of Hub and Axel Bicycle Works.

Figure 10.20
Bickerton collapsible bicycle.

When this is done, the resulting frame is heavier than a diamond frame of equal stiffness. Frequently this is not done, and a flexible frame prone to fatigue failure is the result.

Wheels made from Du Pont Zytel and Kevlar were mentioned briefly earlier and shown in figures 10.7 and 10.8. Most departures from the steel-spoked wheel are made to reduce production cost rather than to improve function. It appears that, as with the diamond frame, experience has evolved a very efficient design having a large margin of safety in ultimate strength (though not in the fatigue strength of the spokes). Articles and arguments abound in bicycling publications concerning spoke-lacing methods and such refinements as rounding the spoke holes in the wheel hub to avoid spoke failures from fatigue in bending (see, for instance, references 14 and 15). Very little scientific work has been published on spoked wheels (although the stresses can be complex[16,17]). This lack of scientific work may be evidence of the few problems associated with this mature design. Such a view is supported by a report in the magazine *Design* by L. Bruce Archer, which occasioned the following commentary in *Cycling*:

Probably the strongest man-made structure relative to its weight is the bicycle wheel. Most high-performance racing wheels are hand-built from factory-made components by craftsmen specializing in this trade. The accepted proportions closely conform to the theoretical optimum, but none of the builders and only a few of the component manufacturers have investigated the matter very deeply. One of the finest British sprinting wheels, weighing only 27 oz., was specially built by Montgomery Young of Condor Cycles, and tested for Design by Dr. B. J. Zaczek, who reported that an axle load of more than 1,200 lbf was sustained before slight inelastic distortion occurred. This is more than

700 times the weight of the wheel. The tyre, inflated to 120 lbf per sq. inch, was compressed flat at the area of contact at an axle load of 400 lbf, but ignoring this the wheel could safely sustain a working load of 700 lbf under smooth riding conditions. This gives the astonishing load/weight ratio for safe working of 400:1, which represents exceptionally high structural efficiency.[18]

References

1. R. E. Peterson, *Stress-Concentration Factors* (New York: Wiley, 1974).

2. J. Awerbuch and H. T. Hahn, Fatigue and proof-testing of unidirectional graphite/epoxy composite, in *Fatigue of Filamentary Composite Materials* (Philadelphia: ASTM, 1977).

3. *Mark's Standard Handbook for Mechanical Engineers*, ed. T. Baumeister, eighth edition (New York: McGraw-Hill, 1978).

4. J. E. Shigley, *Mechanical-Engineering Design*, second edition (New York: McGraw-Hill, 1974).

5. D. J. Kindler, The Design of a Lightweight Tension Wheel with Kevlar Fiber, B.S. thesis, Massachusetts Institute of Technology, 1978.

6. I. Cohen, Polytetrafluoroethylene, *Cycling* (24 March 1955): 301.

7. H. O. Smith, Catalog of the Cycle Collection of the Division of Engineering, bulletin 204, U.S. National Museum (Washington, D.C.: Government Printing Office, 1953).

8. J. Wilde, A cane bicycle from Trieste, *Cycling* (22 December 1945): 420.

9. G. S. Wainwright, Aluminum cycles, *CTC Gazette* (July 1896): 311.

10. A. C. Davison, The Lu-mi-num frame, *Cycling* (19 February 1941): 157.

11. A. Sharp, *Bicycles and Tricycles* (London: Longmans, Green, 1896 / Cambridge, Mass: MIT Press, 1977).

12. *A Shortened History of the Bicycle* (Nottingham: Raleigh Cycle Co., 1975).

13. D. E. Evans, *The Ingenious Mr. Pedersen* (Dursley, U.K.: Alan Sutton, 1979).

14. R. Blum, Avoid spoke and rim problems, *LAW Bulletin* (January 1976): 3–5.

15. R. Jow, Wheels revisited, *Bicycling* (Emmaus, Pa.) (May 1978): 22, 72.

16. A. J. S. Pippard and W. E. Francis, On the theoretical investigation of stresses in a radially spoked wire wheel under loads applied at the rim, *Philosophical Magazine and Journal of Science* (February 1931): 234–285.

17. A. J. S. Pippard and J. White, The stresses in a wire wheel with non-radial loads applied to the rim, *Philosophical Magazine and Journal of Science* (August 1932): 201–232.

18. L. B. Archer, A commentary on wheel design, *Cycling* xx (1956): 362.

Recommended reading

F. DeLong, *DeLong's Guide to Bicycles and Bicycling* (Radnor, Pa.: Chilton, 1974).

F. R. Whitt, Alternatives to metal, *Cycle Touring* (June–July 1971): 138.

———, Bicycles of the future, *Cycle Touring* (August–September 1967): 155–156.

MECHANICS AND MECHANISMS

11

Power transmission

A transmission is the connection between a vehicle's power source and the driving wheel(s). Its purpose is to transmit power with as little loss as possible, and (in the case of bicycles) to transmit it in a way that enables the limbs to move in as near-optimum a manner as possible. In this chapter we review the principles of alternative means of power transmission in bicycles, we examine some examples, and we discuss some possible future developments.

One starting point for this examination is our knowledge of human power generation, which is limited to the circular or linear foot and hand motions used in existing bicycles and ergometers. With the exception of the speed variations given by elliptical chainwheels, the foot velocity in rotary pedaling is a constant proportion of the wheel velocity. Therefore, although we may have hunches that there are other foot, hand, or body motions (or combinations of these) that will enable humans to produce higher levels of maximum power (higher than the upper curves of figures 2.4 and 2.10), or equal levels of power at greater comfort, our scientific knowledge confines us to rotary or linear motions as inputs to power transmissions. For this reason we shall be discussing, principally, rotary pedals and cranks and linear sliders.

To start with, then, we shall limit ourselves to discussing transmissions connecting rotary pedal motions to rotary wheel motions, typified by the familiar pedals and cranks. Let us first make a brief review of the historical development of transmissions to indicate how advances came from perceived needs.

The first "transmission" was linear; to ride a *Draisienne* one pushed one's foot backward on the ground to propel the vehicle forward. The motion was similar to walking and running. However, in walking the legs act as spokes of partial wheels, with the body rolling over the feet, being given both support and propulsion. The essence of von Drais's machine was that the legs were relieved of the need to provide support of the body weight, and could just give thrust. Some downward push was still necessary to provide enough friction, and possibly to maintain balance.

The next two developments were true transmissions that were approximately linear. Louis Gompertz in 1821 added a sector-gear hand drive to the front wheel of a *Draisienne*.[1] This was, no doubt, meant to supplement the foot thrusts, as he provided no footrests. The relatively small amount of power deliverable by the arms, coupled with the need to steer, the evident weight of the vehicle, the solid-rimmed wheels, and the poor road surfaces, must have doomed this design to failure. We have no reports of its use.

Kirkpatrick Macmillan's velocipede, developed between 1839 and 1842, also used an approximately linear drive, with the feet pushing forward on swinging levers (ref. 1). This was the first true transmission, and it enabled the rider to travel long distances with the feet off the ground. Although Macmillan made the rear (driving) wheel larger than the front, it was only about a meter in diameter, and it turned once with every back-and-forth movement of the feet, giving a low gear. However, this probably suited the road conditions of the day. No thread of development followed from Macmillan's pioneering efforts.

Michaux, the first successful developer of rotary crank drive, attached the cranks directly to the front wheel. This was a somewhat simpler arrangement than Macmillan's, and gave the front wheel more freedom to steer, but the wheel diameter was close to that of Macmillan's driving wheel and so a similar low gear was the result.

Michaux was followed by imitators and developers, as Macmillan was not, and the driving wheel was gradually increased in diameter to provide a better coupling, or impedance match, between the human body and the machine. The high-wheeler gave the first combination of a comfortable riding position and an easy rate of pedaling on a two-wheeled vehicle.[2]

This impedance match, or gear ratio, was preserved when chain drive was developed to the extent that a step-up drive between the (separate) cranks and the (rear) driving wheel could be used. The resulting "safety" bicycle was so successful that it is still in essentials the standard bicycle of today.

Thus, by 1885 the principal requirements of a bicycle transmission had been met: to provide a foot motion and a pedaling frequency well suited in average conditions to the capability of the human body to produce power, and to transmit this power from the body (in this case from the feet) to the driving wheel with as little energy loss as possible. The chain drive accomplished both aspects superbly.

Developments to cover nonaverage conditions came fast. A simple approach to low-torque requirements—downhill travel or level running with a strong tailwind—was to fit a one-way clutch or freewheel (figure 11.1) to the chain drive, thus permitting coasting with the feet on the pedals. This removed one possibility of braking but also enabled the rider to "bail out" feet first if necessary.[3]

In high-torque conditions, such as starting, hill climbing, headwinds, or soft ground, riders had to strain at the pedals, often standing on them and pulling up on the handlebars, while pedaling at a very low rate. Scientific testing (see figure 2.18) has confirmed what was intuitively felt: Such pedaling was inefficient. In the twenty years following the introduction of the

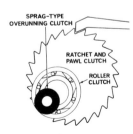

Figure 11.1
One-way clutches.

chain-driven safety bicycle, many different gear-change mechanisms were developed to extend the range of conditions in which a cyclist could pedal efficiently. The two most successful types, the multispeed hub gear and the derailleur gear, have been developed to cover a wide range of conditions and are the predominant types today. In the light of their success, it is perhaps surprising that at the present time there seems to be more invention and development of variable-ratio transmissions using other approaches than of any other aspects of the bicycle.

So much development is going on that to examine more than a few prominent examples of different types would be beyond the bounds of our discussion. Rather, we shall look at some fundamental principles and review alternative possibilities, drawing conclusions where warranted.

Transmission efficiency

Transmission efficiency, η_t, is defined as the energy output at the coupling to the driving wheel divided by the energy input from the human body, usually via the feet. Either energy quantity is measured by the product of force and distance. At a wheel or crank, this can also be expressed as the product of a torque Tq (the force times the radius from the center of rotation at which it acts) and the angle through which it acts, θ (measured in radians). Thus,

$$\eta_t \equiv \frac{Tq_{\text{wheel}} \times \theta_{\text{wheel}}}{Tq_{\text{crank}} \times \theta_{\text{crank}}}.$$

The speed ratio is $(\theta_{\text{wheel}}/\theta_{\text{crank}})$. A perfect transmission, with an efficiency of 100 percent, has therefore a torque ratio $(Tq_{\text{wheel}}/Tq_{\text{crank}})$ that is the inverse of the gear ratio. In practice, meaning with an efficiency of less than 100%, the torque ratio is less than this value:

$$\frac{Tq_{\text{wheel}}}{Tq_{\text{crank}}} = \frac{\eta_t}{\theta_{\text{wheel}}/\theta_{\text{crank}}}.$$

Energy loss in a transmission can occur in two ways. One is friction in bearings and other components. This is the only form of loss in "positive-drive" (chain and gear) transmissions. The other is slip loss, which occurs in transmissions in which the drive is not positive (such as those that use a smooth belt, or some other form of friction or "traction" drive, or an electrical or hydraulic coupling). From this categorization of the forms of energy loss, we can divide transmissions into two broad types: those with and those without positive drive.

Nonpositive drives

It would be easy to dismiss the nonpositive-drive forms from further consideration, because the additional slip losses are a considerable penalty for bicycle application. The chain drive of the first safety bicycles had to compete with the direct wheel-mounted cranks of the high-wheelers, which had an efficiency of almost 100 percent, and any great loss in the chain drive would not have been tolerated. The best roller-chain drives appear to have an efficiency of about 98.5 percent, and the losses would, except in very close races, be imperceptible. But the slip loss of a V-belt, or a hydraulic coupling, or some form of electrical coupling characterized by a generator driving a motor, would multiply these losses by between 5 and 10 unless very large, oversized transmissions were used. The weight and volume of such transmissions would make them unattractive.

There is one possible exception to this almost total elimination of all except positive-drive transmissions: traction drives, or continuously-variable-ratio transmissions.[4] Some well-known types are shown in figure 11.2. The reason these might deserve examination after they have been tried repeatedly and rejected over many years by the major automobile manufacturers is the recent discovery of lubricants that, under high-pressure contact between two hard surfaces, undergo a reversible change in viscosity such

Figure 11.2
Traction drives.

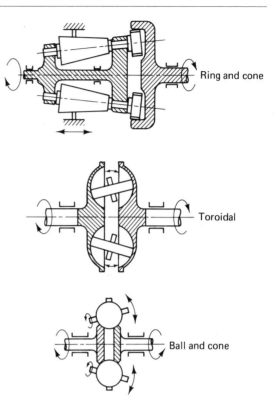

Figure 11.2
Traction drives.

that they can transmit a high shear force.[5] It seems probable that these lubricants will extend the range of usefulness of traction drives outside the very specialized areas to which they are presently confined. However, it is unlikely that any will penetrate the human-powered-vehicle field, because the bicycle requires a transmission that can withstand relatively high torque at low speed in providing a "step-up" speed ratio. Almost all industrial and commercial transmissions have the opposite characteristics in all three respects: low (relative) torque, at high speed, giving a step-down ratio. In fact, the torque capability of a standard bicycle chain drive would enable it to transmit 10–15 kW in an industrial drive (although not with an acceptably long life). Traction drives are already much heavier than their industrial competitors. Therefore, even the use of the new lubricants seems unlikely to overcome their inherent disadvantages for application to bicycles.

Positive drives

Chains and toothed belts

The steel roller chain (in which a freely rotating lubricated roller surrounds each pin) can, together with a front chainwheel and a rear-wheel "cog" or sprocket, constitute a complete transmission. Or the rear sprocket may be attached to a one-way clutch or "freewheel," or to multi-ratio gears (usually enclosed in the rear-wheel hub and incorporating a one-way clutch, as shown in figure 11.3). Or an overlong chain can be used with guiding-plus-tensioner sprockets or "pulleys" that can force the chain to run on one of many in a nest or cluster of sprockets on the wheel and on the chainwheel (figure 11.4).

When new, clean, and well-lubricated, a chain transmission is highly efficient (about 98.5 percent) and very strong (capable of taking the high tension force from a strong, heavy rider ex-

Figure 11.3
Exploded view of Sturmey-Archer five-speed hub gear. Courtesy of T. I. Sturmey-Archer, Ltd.

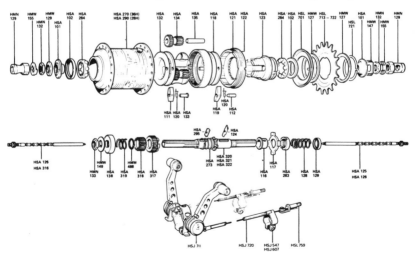

Figure 11.4
Multiexposure
photograph of rear
derailleur during
changing sequence.
Courtesy of Shimano
American Corp.

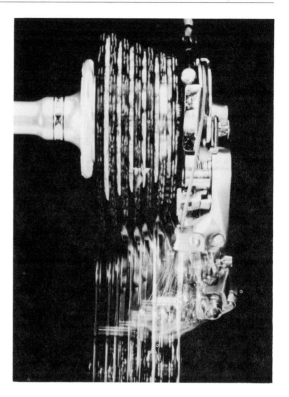

Figure 11.5
Roadster bicycle with
gear case. Courtesy of T.
I. Raleigh, Ltd.

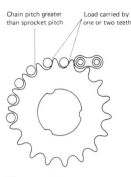

Chain pitch greater   Load carried by
than sprocket pitch   one or two teeth

Figure 11.6
Tooth wear from
stretched chain.

erting maximum force on the pedals). Most bicycles in most parts of the world outside North America and Britain have enclosed chains (so-called gear cases), and their transmissions stay in good condition, often for many years of hard use. (A "roadster" bicycle with enclosed chain drive is shown in figure 11.5.) There is a clear tradeoff of increased weight for higher efficiency, lower maintenance, and longer life. Unfortunately in the opinion of many, chain enclosures have been given an effeminate image in the United States and are no longer available on standard bicycles. The result is that chains, which tend to be in the path of water thrown up by the front tire and of that carried over by the rear tire, often operate in a mixture of old grease, sand and grit, and salt water. Wear is rapid. It is seen as "stretch"; the chain becomes longer, the pitch is slightly larger than that of the chainwheel and the rear cog(s), and the chain therefore tries to ride up the teeth at a larger-than-normal radius (figure 11.6). Instead of the chain load being taken by several teeth, with a stretched chain almost all the load is carried by one tooth at a time, and this further increases wear. A remarkable feature of chain drives is that, even in these very poor conditions, they continue to operate, usually reliably, although the efficiency falls. (We do not know by how much.)

Even more rapid wear is experienced when the chain is used in a multispeed derailleur transmission, for reasons given below. In this case, operation may become unreliable as the worn chain catches on the tips of the worn rear-wheel sprocket teeth and an extra link is periodically carried over, giving a slipping effect.

Let us look at ways in which chain drives could be improved.

A lightweight enclosure of stiff but resilient plastic, such as high-density polyethylene, or polypropylene, or Kevlar-reinforced polyester, should be produced to protect the chain and any derailleur mechanism from dirt, water, snow, and sand.

A smaller pitch (the pin-to-pin spacing) seems desirable to reduce wear and to give a wider choice of ratios. The reasoning is as follows: When a chain wears, it does not, strictly, stretch. No metal is taken even close to its yield point. Rather, the pins wear in the direction of the applied force, and each chain bearing becomes loose. It is this looseness that increases the chain's length.

Mechanical wear is proportional to the product of force and relative movement between two components in contact.[6] Two links in a chain have to undergo relative movement only when they "articulate" onto and off a sprocket. The angle of articulation is equal to $360°/N$, where N is the number of teeth on the sprocket.[7] To obtain the gear ratios which the data of chapter 2 confirm as being desirable, we do not have much freedom of choice for the chainwheel and sprocket diameters. If, however, we halved the chain pitch, from 12.5 mm to 6.25 mm, the angle of articulation of each link would be halved, and the wear would also be halved.

The angle of articulation could be small enough for sliding bearings to be dispensed with in favor of flexing gimbal bearings or flexing links (figure 11.7). A smaller-pitch chain should also be lighter, a factor worthy of consideration because the weight of a chain is relatively quite considerable on a lightweight bicycle. In 1909 the Coventry Cycle Chain Company brought out the "Chainette," a small-pitch (8 mm) chain weighing 1.9 oz/ft (177 g/m), which when tested by *Cycling* was found to run "more like a silken cord" than a chain over sprockets. The British racing cyclist F. H. Grubb broke road records on a bicycle fitted with this chain.

Whether or not chains of smaller pitch were used, friction and wear would be reduced in derailleur gears if "jockey pulleys" of larger di-

Figure 11.7
Flexing gimbal bearing.

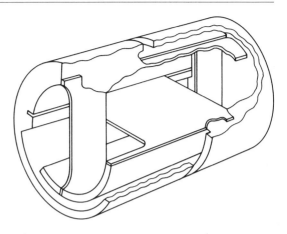

Figure 11.8
SpeeD flexible drive.
Courtesy of Winfred M.
Berg, Inc.

ameter were used. Not only would the angle of articulation be decreased, but the rotation rate of the pulleys, which nowadays usually incorporate plain plastic-to-steel (rather than ball) bearings, would be reduced in the inverse ratio of the increase in diameter.

An attempt to produce a lightweight chain with flexing articulation was made by the Berg Company. Stranded steel cables were used to take the chain tension, and nylon "buttons" took the place of the rollers in a steel chain (figure 11.8).[8] This drive achieved brilliant successes and considerable weight savings in the *Gossamer Condor, Gossamer Albatross,* and *Chrysalis* human-powered airplanes. In this application the driving and driven sprockets were several meters between centers, had rather small step-up ratios, and were at right angles to each other. As yet there has been no successful application to bicycle transmissions because the small diameters of the rear-wheel sprocket and derailleur pulleys have led to fatigue failures of the metal cables (W. Berg, pers. comm.).

The fiber-reinforced toothed belts being used to such a large extent as automobile camshaft drives would seem to be good candidates, at least for bicycles with hub gears. A first attempt to use one by D.G.W. was partially successful. The width required to handle the torque was large: 25 mm. The weight of the toothed belt was much less than that of the chain it replaced, and this more than compensated for the slightly increased weight of the fiberglass-reinforced front sprocket, which was cast onto an existing aluminum chainwheel (figure 11.9). There was some slipping around the rear sprocket at high torque, but this could have been eliminated by a tensioner roller increasing the angle of "wrap" around the rear sprocket. The drive was clean, needed no lubrication, and was unaffected by water or salt. Toothed-belt drives are now being used on Harley-Davidson motorcycles.

Figure 11.9
Toothed-belt drive.

Rather than using several chainwheels and rear sprockets of different diameters and requiring the chain to transfer from one to another, thus introducing misalignment of the chain with the plane of the sprockets, many designers have produced arrangements in which the effective diameters of the chainwheel and/or the sprocket can be changed. Two principal approaches are to use sprockets that are, or become, oval or polygonal. F.R.W. has developed an expanding front chainwheel that, at its smallest (lowest gear), is circular. As a higher gear ratio is called for, two halves of the chainwheel separate and produce greater ovality for the higher gears (figure 11.10). Others use clutched sprockets that move to larger radii along several radial arms

Figure 11.10
Whitt's expanding oval
chainwheel.

Figure 11.11
Cutaway view of
Tokheim transmission,
showing interaction of
"Speedisc" and chain.
Courtesy of Tokheim
Corporation.

Figure 11.12
Hagen all-speed variable-
diameter chainwheel.
Courtesy of Hagen
International, Inc.

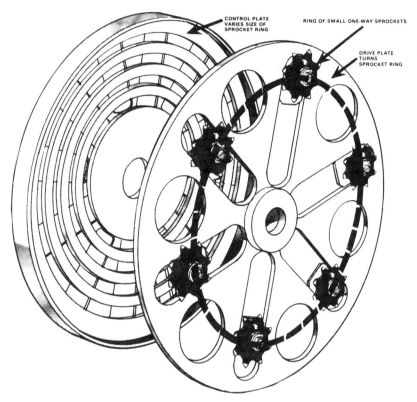

CONTROL PLATE
VARIES SIZE OF
SPROCKET RING

RING OF SMALL ONE-WAY SPROCKETS

DRIVE PLATE
TURNS
SPROCKET RING

Figure 11.13
Octo split-sprocket drive.
Courtesy of Octo Co.

(figures 11.11–11.13). Although we know that the effects on human power output of some degree of ovality of the chainwheel are either negligible or beneficial, we do not know the effects of using polygonal chainwheels or sprockets. We expect these effects to be small. The Octo drive (figure 11.13) uses split sprockets that slide into line with neighboring sprocket segments. This and other similar transmissions allow gear changes under full load. The transmission efficiency of such gears should be slightly higher than that of derailleurs, because they eliminate the small effects of chain misalignment. They promise somewhat easier gear changing than with derailleurs (although there are now available several systems that greatly facilitate derailleur chain shifting). The range of diameters is less, in general, than can be obtained with the wider-range derailleur gears. An automatically self-changing expanding-chain-wheel gear rather similar to that of figure 11.12 has been made by Michael Deal of Bristol University.

Spur-gear systems

Although the word "gear" is used in several different ways in connection with bicycling, in mechanical engineering it refers to toothed spur gears that mesh directly with one another rather than via a chain or toothed belt.

When a set of gears is designed to give a speed (or torque) ratio between input and output shafts, two alternative approaches are possible. In one, all the axes around which the individual gears rotate are fixed relative to the casing (figure 11.14); in the other some of the gear axes themselves rotate around a center (figure 11.15). The latter are called epicyclic gears. Virtually all bicycle spur-gear systems used at present are epicyclic, principally because of the compact ar-

Figure 11.14
Fixed-axes gears. If the gear on shaft A has T_A teeth and that on shaft B has T_B teeth, then one turn clockwise $(+1)$ of shaft A will turn shaft B counterclockwise $-(T_A/T_B)$.

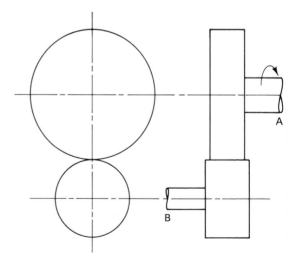

Figure 11.15
Moving-axes (epicyclic)
gears. Inputs and/or
outputs can be connected
to A, C, and D. In a
bicycle hub gear, A is on
a stationary shaft. In the
low gear, the chain-
sprocket input is
connected to C and the
output (D) is connected
to the wheel hub. In top
gear, these connections
are reversed. The gear set
is bypassed in middle
gear, with the sprocket
connected via the
freewheel to the wheel
hub.

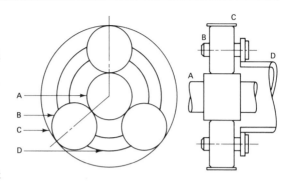

rangement that is possible. Though at different
times gear-change systems have been developed
to fit the bottom-bracket or crank position, these
have tended to be large because they must withstand the full cranking torque. In the rear-wheel
hub, connected to the chainwheel by a conventional chain, the torque is reduced by the chain-
wheel-to-sprocket ratio (usually about 3:1), so
that a hub gear can be designed to one-third the
peak torque of a bottom-bracket gear.

The calculation of speed ratios is illustrated by
table 11.1. The design of any spur-gear transmission is highly specialized. Standard mechan-
ical-design texts provide excellent guidance, but
they are usually written for industrial applications, for which machines may sometimes be
expected to operate for 100,000 hours. A bicycle
(or an automobile) has a relatively short operating life (1,000–2,000 hours), and their transmissions were developed to their present compact
sizes and configurations through early intense
efforts to reduce weight, volume, and cost.

The Sturmey-Archer five-speed hub gear
shown in figure 11.3 incorporates two epicyclic
gear sets. The five speeds are given by direct
drive; by input to the "cages" of the "planets"
of either gear set, with output from the appropriate ring gears (in the step-up gears); and by
inputs to the ring gears in turn with outputs
from the plantetary cages. Ingenious systems of

Table 11.1 Calculation of ratios in an epicyclic gear set. Refer to figure 11.15.

Step	A (shaft and pinion)	B (gear)	C (ring)	D (cage)
Stop rotation of D; turn shaft A -1	-1	$+ (T_A/T_B)$	$+ (T_A/T_C)$	0
Fix all gears relative to each other and rotate whole gear set $+1$	$+1$	$+1$	$+1$	$+1$
Resulting ratio (add)	0	$1+T_A/T_B$	$1+T_A/T_C$	$+1$

If $T_A = T_B$, then the top-gear ratio is 1.333 and the low gear is 0.75 (because $T_C = T_A + 2T_B = 3T_A$).

Figure 11.16
Efficiency of epicyclic
hub gears. Curves from
data in reference 9;
points determined
experimentally by F.R.W.

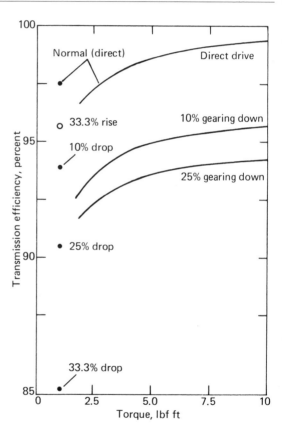

sliding "dog" clutches and one-way clutches
are used to give these changes with two cable-
operated levers or chains, one from each end of
the hollow main spindle. The original three-
speed versions of these gears had reached virtually their present form by 1909 (when ball-
bearing mounting of the planetary gears was
abandoned). The individual gears and many of
the other components are now made by powder
metallurgy, which gives high accuracy without
machining.

The efficiency of epicyclic hub gears as measured as a function of applied torque[9] is illustrated in figure 11.16.

Shaft drives

Some early safety bicycles used shaft drive in
place of a chain, with a right-angle bevel-gear
set at the crank and at the rear wheel (figure
11.17). These drives had a neat, compact appearance, but were heavier, less efficient, and
much more expensive than chain drive.

In the period around 1897, most American
manufacturers produced at least one shaft-drive
model. Tests showed that the losses could be as

Figure 11.17
Pierce shaft-drive bicycle
with sprung rear wheel,
1900. Courtesy of
Smithsonian Institution.

high as 8 percent, probably because the machining of bevel gears was not very precise. The Waltham Orient pattern using roller pins instead of machined teeth performed well, however, and Major Taylor broke many records using this transmission.

Linear and oscillating transmissions

An early form of linear transmission is illustrated in figure 2.13. When the pedal is pushed there is no resistance until the pedal velocity has caught up with the wheel velocity at the setting of the gear-ratio adjustment on the radius arm. For torque to be transmitted smoothly at this point, it is essential that a one-way clutch without backlash or overshoot be used. Such a device is the sprag clutch, which is shown in figure 11.18.

The linear or oscillating drive has attracted many because it can give a continuously-variable-ratio transmission, which is apparently well matched to a natural ladder-climbing action of the legs. However, above very low "pedaling" speeds the energy required to speed the legs and feet up to the speed of the wheel is a considerable loss. Muscle energy must also be used wastefully to slow down the legs and feet at the end of the stroke. The Ball tricycle *Dragonfly II* with a single prone rider (Rick Byrne) and with hand-and-foot linear drive (figure 11.19) achieved 54.7 mph (24.4 m/sec) for second place in the 200-m flying-start IHPVA speed trials in California in May 1980, showing that the losses could not have been large.

Several methods of combining the advantages of the oscillating drive and its wide choice of gear ratios with an energy-conserving pedaling system have been developed. The most recent is the Bio-Cam, shown in figure 2.14. In this case the designer has considered that rotary pedaling motion, which of course conserves kinetic energy, is more ergonomically efficient than a lin-

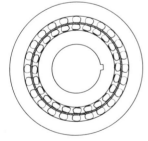

Figure 11.18
Sprag clutch.

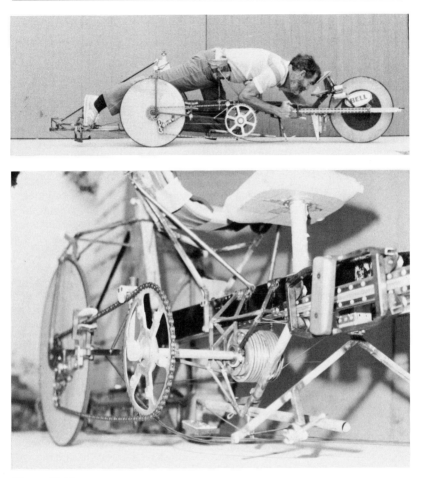

Figure 11.19
Steve Ball's linear hand-
and-foot drive. Courtesy
of Steve Ball.

ear motion. The cam on the crankshaft operates a form of oscillating drive.

Hydrostatic drive

Heavy earth-moving equipment, for example, often uses a type of transmission in which the input runs a positive-displacement hydraulic pump and the high-pressure oil is piped to a hydraulic motor that drives the output shaft. A major advantage is that a type of variable-angle-swashplate axial-piston pump permits the output to be varied over a wide range from positive to negative flow, giving a continuous variation of speed ratio. There have been many attempts to apply this transmission to passenger automobiles and to bicycles. An apparently insurmountable problem is that the peak efficiency of a hydraulic pump and a hydraulic motor is about 90 percent, so the overall transmission efficiency cannot be much over 80 percent. This is unacceptably low for a human-power application.

Conclusions

The efficiencies of present transmissions using chains and derailleurs or hub gears are in the high nineties, and any future improvements must be small. There is scope for weight reduction, for protection against deterioration, and for some input motions of the feet alone or of the feet plus the hands that would allow a higher maximum power to be delivered than with existing circular constant-velocity pedaling. It is presumed that such an alternative input motion could give similar power levels with greater comfort.

References

1. A. Ritchie, *King of the Road* (Berkeley, Calif.: Ten-Speed, 1975).

2. A. Sharp, *Bicycles and Tricycles* (London: Longmans, Green, 1896 / Cambridge, Mass.: MIT Press, 1977).

3. V. Bury and G. L. Hillier (eds.), *Cycling* (London: Longmans, Green, 1887).

4. R. W. Carson, New and better traction drives are here, *Machine Design* (18 April 1974): 148–155.

5. R. L. Green and F. L. Langenfeld, Lubricants and traction drives, *Machine Design* (2 May 1974): 108–113.

6. F. A. McClintock and A. S. Argon, *Mechanical Behavior of Materials* (Reading, Mass.: Addison-Wesley, 1966).

7. J. E. Shigley, *Mechanical Engineering Design,* second edition (New York: McGraw-Hill, 1972).

8. Precision Mechanical Components (catalog), Winfred M. Berg, Inc., East Rockaway, N.Y., 1979.

9. A. Thom, P. G. Lund, and J. D. Todd, Efficiency of three-speed bicycle gears, *Engineering* (London) 180 (2 July 1956): 78–79.

Recommended reading   F. R. Whitt, Variable gears: Some basic ergonomics and mechanics, in *Developing Pedal Power* (Milton Keynes, U.K.: New Towns Study Unit, Open University, 1979).

12 Unusual pedaled machines

Off-road vehicles

As stated before, it is probable that widespread development of better roads made the use of bicycles much more practical. The propulsive power needed was then brought below that for walking or running at comparable rates, and the encumbrance of a machine became justifiable. Although walking on soft ground requires twice as much effort as walking on concrete, a wheeled machine on soft ground experiences about a fiftyfold increase in resistance. Thus, on soil the advantages of a wheeled machine over a walker are diminished.

Most of the world's roads are made of bonded earth, with relatively poor surfaces, and as a consequence bicycling in general is under less-than-optimum conditions. The bicycles used on these roads are somewhat different from those now used on good American and European roads. Throughout the rest of the world—and particularly where roads are poor—the 28-inch (about 700 mm) wheel with a tire about $1\frac{1}{2}$ inches (39 mm) in cross section is commonly used. Big wheels with large tires have also provided a partial solution for agricultural and military vehicles, which have to travel on poor surfaces.

In addition to attempting to solve the problems associated with the use of vehicles on poor roads, inventors have tried to devise human-powered vehicles for other situations. Riding on water, on railways, and in the air have been goals of inventors ever since the practical bicycle appeared and demonstrated its speed on good roads in the late nineteenth century.[1-4] It is probable that the bicycle's high efficiency under good conditions was taken mistakenly by many inventors to imply that similar performances could be expected from human-powered vehicles under very different conditions.

Boats

By building harder, smoother roads, mankind became able to use wheeled machines to great advantage in traveling with a minimum of effort. However, it is not possible to duplicate this achievement by producing smoother water. The resistance to movement offered by a relatively dense and viscous medium such as water is great compared with that offered by air. As a consequence, both submerged and floating objects (such as swimmers and rowboats) can travel at only a quarter of the speeds reached with similar efforts by their land counterparts (runners and bicyclists).

Rather surprisingly, there has been no pattern of progressive innovation in human-powered water vehicles since the development of the modern racing shell. The body is capable of giving out most of its energy through the leg muscles, yet throughout history most human muscle power has been applied through the arms and back. Moreover, these muscles have usually been worked at totally unsuitable rates of application. That is the case in the slow pushing or pulling of oars in a fixed-seat rowboat; arms are better suited to rapid cranking. The sliding-seat racing shell was, therefore, a major development.

There are two principal differences between a conventional rowboat and a racing shell. One is the length-to-beam ratio, which in shells is made as high as 50:1 to give minimum combined wave and skin-friction drag. The other is that the rower's seat slides forward and aft, permitting the rower's energy to be delivered principally by the leg muscles instead of by the arms and back as in a rowboat.

Rowing is not the most efficient propulsion method. Reciprocating oars, sculls, and paddles waste energy. The relative kinetic energy of the boat, the oars, and the rower or paddler are reversed every half-cycle by muscle energy. During the return stroke, oars cause high wind drag. During much of their movement through the water oars dissipate energy into side thrust, which must be resisted by opposing oars or by rudders and fins, introducing more drag; the skilled canoeist's J-stroke produces similar advantages and disadvantages. Rowing is also somewhat hazardous; if because of misjudgment or an unexpected wave an oar comes out of the water when the rower is pulling hard, or catches the water unexpectedly during its return, there is a risk of upset. And oars and paddles are inconvenient; one's hands are fully occupied while underway, so one cannot easily hold a rudder or tiller, a fishing rod, or lunch.

Most human-powered boats of all types have been powered by oars and paddles, a few by paddlewheels, even fewer by poling, and an almost negligible number by screws. Reference 5, among other publications, states that oars, paddlewheels, and screw propellers all have efficiencies of about 70 percent. However, this statement by itself is somewhat misleading for at least two reasons. One is that the efficiency it refers to is that under design-point conditions. Under other conditions the efficiencies of paddlewheels and oars fall off more rapidly than that of propellers. In 1845, the British Admiralty staged a tug of war between two ships of similar displacement and installed power, one with twin paddlewheels and the other (H.M.S. *Rattler*, designed and built by Isambard Kingdom Brunel) with a screw.[6] The *Rattler* towed the paddlewheel steamer astern at 2.5 knots. There was, therefore, no doubt about the superior efficiency of screws in low-speed high-power conditions. The second reason why the three alternative methods of propulsion are not equal in efficiency is that there have been substantial developments in the fluid mechanics of propellers, both for air and for water. A design method recently refined by Larrabee[7] was used for the first human-powered English Channel flight, in the MacCready team's *Gossamer Albatross*. The pilot, Bryan Allen, reported that when the team switched to the Larrabee-designed propeller, his endurance increased from a maximum of 10 minutes to one limited mainly by the need to land to take on water and food. The propeller's efficiency was estimated to be about 87 percent.

Even without the modern improvements in propeller efficiency, evidence of the superiority of screws over oars is provided by reference 1's account of the performances of water cycles in their heyday of the 1890s. A triplet water cycle (figure 12.1) ridden by the former racing bicyclist F. Cooper and two others covered 101 miles (162.5 km) on the Thames from Oxford to Putney in 19 hours, 27 minutes, and 50 seconds. A triple-sculls boat rowed by good university oarsmen covered the same course in 22 hours and 28 seconds. The water cycle was the faster vehicle by about 18 percent.

Other facts about water cycles propelled by pedal-driven screws in this period are interesting. The English Channel was crossed (from Dover to Calais) by a tandem water cycle in $7\frac{1}{4}$ hours. A sextuple water cycle ridden by women on the Seine is credited with reaching a speed of 15 mph (6.7 m/sec). "Hydrocycles" manufactured by L. U. Moulton of Michigan were said to be capable of 10 mph (4.47 m/sec). All these performances compare favorably with those of boats rowed by the best oarsmen.

Figure 12.1
Triplet water cycle.
Courtesy of Currys, Ltd.

Two applications of screw propulsion could be especially appropriate and fruitful: utility use and sport. A utility boat is one used for ferrying to a boat out at mooring, or for a day's fishing on a harbor or a lake, or for fetching groceries for an island home from the mainland shore. All of these duties used to be served by conventional rowboats, but nowadays many people use outboard-engined boats, which consume gasoline and oil and are often noisy, polluting, and unreliable. Only modest development is needed to produce an efficient outboard screw drive that could be swiveled to provide high maneuverability[8] (figure 12.2). Figure 12.3 shows a design for a high-speed sport or recreation boat. In the 1890s there were twin-hulled racers with two pedalers mounted centrally in conventional tandem bicycling position. Brewster studied this and two alternative configurations:[9] one with a single submerged low-drag hull for most of the buoyancy and outrigger pontoons for stability, and one with hydrofoils. On paper the hydrofoil boat is the faster design, able to attain 10 mph (4.5 m/sec) with a power input of about 200 watts (figure 12.4). Atheletes such as racing cyclists capable of producing 375 watts for an hour or more could, if the analysis is correct, propel the design to over 17 mph (7.5 m/sec).

"Amphibious" machines have been constructed and ridden. These had floats arranged so that when the machine was ridden on land they did not obstruct its movement (refs. 2, 3).

Data on the drag of hulls, submerged bodies, and foils are given in reference 10. An approximate relation among power, area of wetted surface, and speed of streamlined hulls is

$$\text{hp} = 2.4 \times 10^{-5} \times \text{Wetted surface (ft}^2) \times \text{knots}^{2.86},$$

or

$$\text{watts} = 1.287 \times \text{Wetted surface (m}^2) \times \text{(m/sec)}^{2.86}.$$

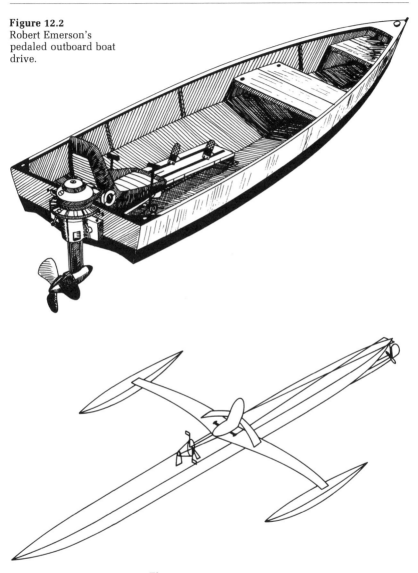

Figure 12.2
Robert Emerson's
pedaled outboard boat
drive.

Figure 12.3
Design for high-speed
human-powered boat.

Figure 12.4
Power requirements for
propulsion of high-speed
boats. Adapted from
reference 9.

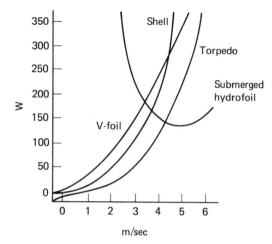

Further development of human-powered boats
is likely to come from the incentives of amateur
competition. The IHPVA started a water-craft
division in 1981. Improvements as dramatic as
those with road and track vehicles can be expected. The Texas Water Safari, a race of over
400 miles in a range of different water conditions, has an "unlimited" class and will also
spawn useful developments in human-powered
boats.

**Ice and snow
machines**

There have been attempts to develop and popu-
larize bicycle-type machines for running on ice
or snow (ref. 3). Some types consist of a bicycle
with a ski replacing the front wheel; others dispense with wheels and retain only the bicycle
frame, with a ski attached on either side. In
contrast with the case of water cycles, there
have been no published comparisons of the
speeds of these machines with those of skaters
or skiers.

Railway cycles

The coefficient of rolling resistance of a steel
wheel running on a steel rail is in the range of
one-tenth of that of the best of pneumatic-tired
wheels running under optimum conditions of

Figure 12.5
Early railway cycle.
Courtesy of Currys, Ltd.

road use. As a consequence, cycles developed for running on rails have been proved practical in the sense that they were not difficult to propel. In fact, high speeds are credited to this type of machine. An illustration of one type is given in figure 12.5 (see also references 2–4). A simpler approach was used by Bob Delgatty of Victoria, B.C., who joined two standard independently ridden bicycles together with a tubular structure. In this case the regular pneumatic tires run on the railheads, and side wheels maintain lateral direction. Steering is, therefore, not required, and Delgatty reports delight in being able to photograph and to enjoy the view freely while pedaling. Other designs are shown in reference 11.

A drawback to railway cycles was the general unavailability of unused lines. The Victorians took quite seriously the idea of laying special cycle tracks alongside the regular rail tracks in some areas. With the abandonment of many branch lines, there have been calls to preserve them for this kind of recreational use.

Aircraft

There have been attempts, both serious and maniacal, at unaided human-powered flight since at least 1400 B.C. Da Vinci pursued this dream in many sketches, but it proved elusive. Only two celebrated aircraft designed and built by a group led by Paul MacCready of Aero-Vironment, Inc., in Pasadena, California, have achieved great success, winning prizes totaling £150,000. The purse was proffered in two challenging contests by Henry Kremer, a British industrialist. In 1959 he offered £5,000 (about $20,000 at the time) to the first group to fly a human-powered aircraft at least 10 feet from the ground through a figure-eight course around two pylons placed a half-mile apart; he later raised the prize to £50,000. Team after team produced planes that could fly in straight lines and make small banking curves, but failed to negotiate the course. In 1977 MacCready's *Gossamer Condor* succeeded for three reasons: because it incorporated sophisticated aerodynamics (as did many of the others); because it was designed to cruise at only 9 mph (3.6 m/sec) and thus required less power than many of its competitors, most of which were designed for 13–20 mph (6–9 m/sec); and because the airframe had been designed to be rebuilt in a matter of hours or even minutes after a crash. By rigorously testing each version of the design, analyzing errors, and rebuilding quickly, MacCready's group was able to accomplish in one year what others had failed to do for 18 years.[12]

Henry Kremer then put up £100,000 ($180,000) for a human-powered flight over the English Channel. This time the prize went unwon for barely one year. In June 1979 MacCready's *Gossamer Albatross* (figure 12.6), weighing under 60 lb, went the distance on its first attempt, with Bryan Allen again serving as pilot and power plant. Allen produced about $\frac{1}{3}$ hp (250 W) for about 3 hours, whereas Blériot's monoplane made the first flight over the Channel in 36 minutes using a 25-hp (19-kW) engine.

The maximum wind speed in which such an airplane can be flown is only around 2–4 mph (1–2 m/sec)—practically a dead calm—because ground-level air turbulence has vertical components approaching wind speed. A modest gust

Figure 12.6
The *Gossamer Albatross*.
Courtesy of E. I. du Pont
de Nemours Co.

of 9–11 mph may severely damage a flimsy, low-speed aircraft even if it is held down and not allowed to become a kite, which it would want to do. Such periods of calm typically occur only for an hour or two near dawn on very still days. As a result these aircraft cannot safely go far from their large and expensive hangars, in which they must spend most of their time.

There will be small improvements in human-powered heavier-than-air flight, but because of physical constraints such flight must be limited to the areas of sport and recreation. Improvements will be slow in coming because physical limits are being approached. For example, at under 60 lb (27 kg) the *Gossamer Albatross* is already a fraction of the weight of the pilot, and further weight reductions will not bring proportional benefits.

For the present, human-powered flight must be regarded as an expensive prospect with great limitations. But improvements and innovations could change that. Foldable planes might eliminate hangar costs. Some lift from helium, hot air, or possibly hydrogen would reduce the required lifting power and allow more power to be devoted to propulsion. An inflatable airplane is a possibility. Furthermore, the low gravity of large space colonies, say one-sixth to one-tenth that of earth's, could make high-speed human-powered transportation in the air practicable there.[13]

Lawn mowers

Figure 12.7
Diagrams of the Shakespear pedaled mower. From reference 14.

The rationale behind the design of the mower shown in figures 12.7 and 12.8 was that the leg muscles would be used more efficiently in pedaling than in pushing a conventional mower, and the back and arm muscles would be relieved; that continuous mowing would be more efficient than the frequently used to-and-fro motion of push mowing; that a multiratio gear would enable users to choose the power-output

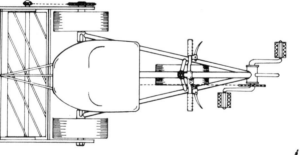

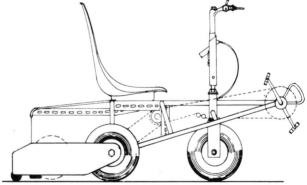

Figure 12.8
Michael Shakespear on
his mower.

rate and make it easier to mow on moderate
slopes; that gasoline shortages and antinoise
regulations might limit the use of power mowers; and that riding a pedaled mower might be
fun as well as good exercise.

The original model shown in the figures was
designed and constructed by Michael Shakespear of the Massachusetts Institute of Technology for his bachelor's thesis in mechanical
engineering.[14] A three-speed Sturmey-Archer
hub gear, a brake, and a differential are incorporated into the transmission. The reel-type cutter
is driven directly from the input to the differential drive to the rear wheels. Pulling the left
handlebar lever releases a catch and makes it
possible to raise the cutter assembly by pulling
the handlebars back; this allows easy maneuvering. The prototype, built largely of scrap materials and components, was very heavy but still
gave easy cutting. A lightweight model might

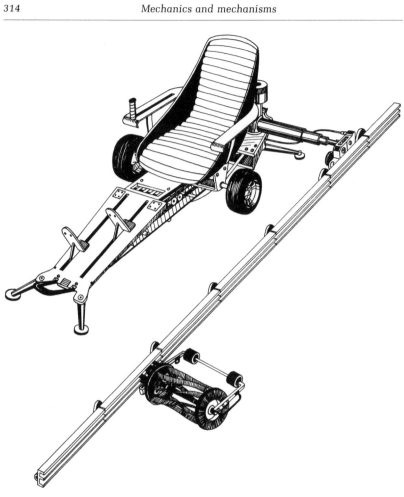

Figure 12.9
Human-powered
''satellite'' mower.

show real advantages. A more sophisticated approach that would reduce the large energy losses of wheels on soft ground would be to have a stationary power unit and a "satellite" cutting mechanism (see figure 12.9 and reference 11).

Energy-storage bicycles

The concept of storing braking or downhill energy (or even energy pedaled into the machine while the rider waits at a traffic light), and then drawing on the stored energy for a burst of power to accelerate or to climb a hill, has intrigued inventors for many years. In every bicyclist there is a suppressed desire to leave sports cars standing in a cloud of rubber smoke. Sad to say, the chances are small.

Table 12.1 gives the maximum energy-storage capabilities of various systems.[15] Flywheels are so much better than rubber bands or springs that they would be the preferred contenders, and they have many enthusiasts.[16] (City buses driven by flywheels were manufactured by Sulzer in Switzerland; the flywheels were sped up by electric motors at stops. This system is being given much new attention.) Compared with the energy-storage capability of gasoline, however, a flywheel is almost 100 times heavier. Also, a flywheel needs a continuously-variable-ratio transmission if its kinetic energy is to be transferred efficiently to the driving wheel. And the "windage" losses constantly degrade the stored energy. All these factors mean high weights and high losses, neither welcome in bicycling.

Batteries are better as far as the weight of the energy storage alone is concerned, but then a motor, a control system, and a transmission are required. At least 0.5 hp (400 W) would be desirable, and a minimum weight for a special motor and transmission might be 10 lb (4 kg). The battery and its housing would be another 10 lb. (Extremely expensive aerospace-type components would be required to keep weights

Table 12.1 Energy-storage data.

Maximum energy-storage capability of various materials

Material	Electrochemical conversion[a]		Heat-engine conversion[b]		Mechanical conversion	
	W-hr/lbm	kJ/kg	W-hr/lbm	kJ/kg	W-hr/lbm	kJ/kg
Hydrogen[c]	14,900	118,250	3,040	24,130		
Gasoline[c]	5,850	46,430	1,130	8,970		
Methanol[c]	2,760	21,900	505	4,010		
Ammonia[c]	2,520	20,000	503	4,000		
Hydrogen-oxygen (liquid)	1,660	13,175	338	2,680		
Lithium-chlorine (700°C)	980	7,780				
Magnesium-oxygen[d]	1,800	14,290				
Sodium-oxygen[d]	775	6,150				
Zinc-oxygen[d]	500	3,970				
Sodium-sulfur (300°C)	385	3,060				
Lithium-copper-fluoride	746	5,920				
Zinc-silver dioxide (silver-zinc-battery)	208	1,650				
Lead-lead dioxide (lead-acid battery)	85	675				
Cooling lithium hydride			64	508		
Flywheel					14	111
Compressed gas and container					10	79
Rubber bands					1	8
Springs					0.06	0.5
Capacitors					0.006	0.05

Energy density

System	Low		High	
	kWh/ft$_3$	kWh/m$_3$	kWh/ft$_3$	kWh/m$_3$
Electrostatic			0.0045	0.16
Magnetic	0.0007	0.025	0.06	2.1
Gravitational	0.006	0.21	0.15	5.3
Mechanical	0.0007	0.025	0.6	21.2
Phase change	0.007	0.25	75.	2,650
Primary battery	0.15	5.3	7.5	265
Secondary battery	0.45	15.9	1.5	53
Fuel cell	0.75	26.5	75.	2,650
Fuel			300.	10,590

Source: reference 15, p. 54.
a. Based on Gibbs free energy
b. Assumes 20 percent thermal efficiency
c. Reaction with oxygen from atmosphere
d. Including weight of oxygen

down to these levels.) A lightweight bicycle would about double its weight, and the rider might well think about going a step farther to a motorcycle (maybe even a battery-powered one).

These conclusions have been given some weight by a study performed by students at Dartmouth College.[17] They adopted specifications that included a price of fifty 1962 dollars, a weight of 30 lb (13.6 kg), and a power output sufficient to propel the rider and the machine up a hill 2,120 ft (645 m) long and 90 ft (27.4 m) high. Four systems were studied: a spring, a flywheel, electrical storage, and hydraulic storage. It was decided that there was no spring system that could be described as practical. The hydraulic system would have cost $1,500 and would have been heavy because it would have had to work at extremely high pressures. The mechanical flywheel system considered would have been suitable if it incorporated two 35-lb (15.9-kg) flywheels revolving at 4,800 rpm; these characteristics made the concept too expensive. The electrical system, consisting of a motor-generator and an electricity accumulator, would have had a price of $74, an overall efficiency of 34 percent, and a weight of 40 lb (18.1 kg). This is much nearer to the specification. However, the low efficiency and the high weight and cost made the concept very unattractive.

Figure 12.10
Thompson flywheel bicycle.

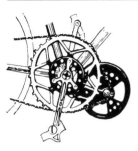

Figure 12.11
Flywheel system for
racing bicycle.

For a flywheel to have the maximum energy-
storage capability per unit mass, the rim must
be moving at the highest velocity that stress
considerations will allow. For this reason, the
approach used in the Thompson flywheel bicycles (figure 12.10) is not valid. The geared-up
system of figure 12.11 is a little better, but still
unattractive. Recent developments incorporating
flywheels of composite materials rotating in a
vacuum with sophisticated electrically modu-
lated transmissions cannot yet be considered for
bicycles.

For further discussions of energy storage, including earlier attempts, see references 3 and
18.

Cyclecars

We are using this term to distinguish three- and
four-wheeled pedaled vehicles with automobile-
like bodies from the bodyless tricycles and
quadracycles that followed more traditional bi-
cycle-construction practice from 1880 on. There
have been such a large number and a wide variety of cyclecars made in many countries that all
we can do here is to refer briefly to those we
believe to be the most worthy of note.

We mentioned in chapter 1 that the Velocar recumbent bicycle was named after a similarly
propelled cyclecar. Cyclecar racing had become
popular in the 1920s, particularly in Germany.
The principal proponent was Manfred Curry,
principally known for his research and his text
on fast sailboats. He designed and made the
Landskiff ("land boat") (figure 12.12), known in
the English-speaking countries as a Rowmobile
because the driver used the same motions as in
rowing a sliding-seat shell. (A similar principle
had been used by George W. Lee in the United
States for a tricycle probably constructed in the
1870s; see figure 12.13.) The Curry machines
were used for racing and for commuting, and
(according to a 1930 catalog for the "new improved model" made by Alexander Metz of
Munich[19]) weighed 35 kg in the single-rider

Figure 12.12
Curry *Landskiff* (1930)
for driver and passenger.

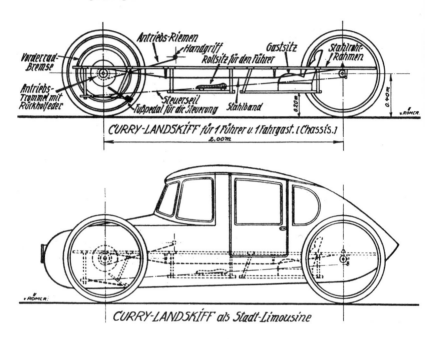

CURRY-LANDSKIFF für 1 Führer u. 1 Fahrgast. (Chassis.)

CURRY-LANDSKIFF als Stadt-Limousine

Figure 12.13
Rowing-action tricycle.

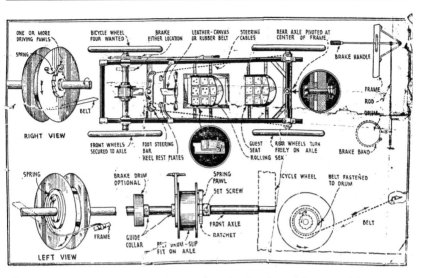

Figure 12.14
Diagram from "How to build a rowmobile," *Science and Invention* (August 1926).

version and 45 kg for the tandem machine. They were well streamlined, and were reputed to reach 30 mph (13.5 m/sec) on smooth, level surfaces. Instructions on building a rowmobile (figure 12.14) appeared in several mechanics' magazines in the 1920s and 1930s.[20,21]

Several "people-powered vehicles" appeared in the United States in the early 1970s, and one (figure 12.15) was commercially produced in numbers. Although they caught the imagination of newspeople, they did not displace either automobiles or bicycles on the highways or, to any great extent, tricycles in retirement communities. Some of the reasons for this lack of success are discussed in the next chapter.

A compromise between the almost totally enclosed cyclecars with their rigid bodies and the completely open standard tricycles was the range of "all-weather cycles" designed in the 1970s by Paul Schöndorf in Cologne. The "Easy Muscar" shown in figure 12.16 retains the standard tricycle's virtues of easy access, comfort, luggage-carrying ability, and some weather protection without a large weight or size penalty.

Figure 12.15
PPV ("people-powered
vehicle").

Figure 12.16
Schöndorf's "easy
Muscar" tricycle.

Human-powered vehicles in the Third World

Where the automobile is economically out of reach for most people—over much of Africa and Asia and some of Europe—the bicycle still reigns supreme. The Viet Cong were supplied by trains of bicycles, and when D.G.W. lived in Nigeria in the late 1950s a bicycle was a highly prized possession, often taking precedence over a wife (whose purchase price was often comparable). Bicycles are the mass-transit system in much of the Third World. Pedal-powered vehicles are also used as short-distance taxis and freight carriers. Considerable ingenuity is often used to adapt bicycle components to three- and four-wheeled vehicles, although there is little technical sophistication.

In African countries, bicycles are used for both personal and freight transportation. Goods (for instance, two or three mattresses) will often be carried on the head, gracefully balanced. In Nigeria in 1958–1960, the almost universal machine was the 28-inch-wheel Raleigh three-speed with a gear case around the chain. Often the shift trigger was bound permanently to the handlebar with copper wire in the low-gear position. Mirrors were fixed in the middle of the handlebar to check on the rider's appearance rather than to view traffic coming from the rear. Bicycles arrived from Raleigh with the frames spiral-wrapped in brown paper, and a vigorous rewrapping industry arose to serve those who wanted to maintain the new-bicycle appearance. These observations are made not so much to be condescending as to point out that in an early stage of development, people of any country will use Western technology in an unquestioning and distinctive manner.

In India and most of Asia, three-wheeled vehicles are used widely. Reference 22 points out that there are three basic load-carrying designs—driver in front, driver behind, and sidecar:

*In the driver-in-front tricycle, steering remains
as on a regular bicycle, with power transmitted
via a long chain to the two rear wheels. The
driver-in-rear configuration features two front
wheels, with power supplied to a single rear
wheel. Steering is accomplished by turning the
entire front compartment.*

*Each design has its own advantages and
disadvantages. The driver-in-front design, used
mostly in India, is lighter and easier to pedal
and steer. However, side-car and driver-in-rear
designs, found in Indonesia, Malaysia, and
Vietnam, can carry heavier loads, because
power is transmitted more directly.*

(It is not clear what is meant by "more
directly," but it is probably that usually no differential is used and that when two wheels are
at the rear only one is driven.) Only rarely are
hub gears used, and the single gear ratio is
chosen as a compromise to allow heavy loads
(perhaps 330 lb, or 150 kg) to be started from
rest and taken up gentle inclines while permitting cruising speeds of about 4–7 mph (2–3
m/sec). There is little attempt to save weight;
when heavy loads are being transported, any
saving in frame weight that might lead to early
failure is a poor tradeoff. According to a personal communication to D.G.W. from M. M. Negru of Kerala, South India,

*. . . the drivers are hard on the cycles since they
are rented rather than owned; the machine must
be able to stand this. The roads are clogged
with trucks, buses, cars, scooters, buffalo,
bullock carts, goats, pigs, assorted pedestrians,
ordinary bicycles and the odd temple elephant.
With such a mix of speeds, traffic is chaotic and
the bicycle rickshaw must be at least as
manoeuvrable as current models; fast swerves
average two per km. . . .*

Negru also states that these rickshaws "are
clumsy inefficiently designed contrivances,

which are losing their passengers to the faster, more comfortable 'autorickshaws' (a golf-cart-type affair with a scooter motor carrying three or four persons including the driver)." He calls for improved designs of human-powered vehicles.

Groups such as the Intermediate Technology Development Group (U.K.) and Volunteers In Technical Assistance (U.S.) have tried to infuse some new technology. In particular, S. S. Wilson of Oxford University, with the support of the relief agency Oxfam, has developed the load-carrying Oxtrike (figure 12.17), which is made from simple components and suitable for indigenous manufacture but incorporates a three-speed gear. In reference 11 S. S. Wilson gives details of the construction of the Oxtrike and of other human-powered devices. (That reference also contains a wide variety of other pedal-powered machines.)

Figure 12.17
Oxtrike. Courtesy of S. S. Wilson.

It seems surprising that the bicycle trailer, which has been used for years in homemade versions in Europe and has been produced in

several lightweight commercial versions in the United States in the 1970s, should not find application in developing countries. A trailer allows an unmodified bicycle to be used for load hauling without adding greatly to the stresses it has to withstand. The trailer is hitched to the saddle stem or to the rear-wheel nut, and interferes remarkably little with the handling. However, braking deteriorates considerably, and trailers loaded with goods or children should be used with great caution.

References

1. *The Rambler* (London: Temple House, 1897).

2. *Strange but True,* nos. 12, 26, and 48, Curry's Ltd.

3. A. J. Palmer, *Riding High: The Story of the Bicycle* (New York: Dutton, 1956).

4. J. Hadfield, *Saturday Book* (Boston: Little, Brown, 1965).

5. *A Dictionary of Applied Physics,* ed. R. Glazebrook (London: Macmillan, 1922).

6. *The New Encyclopaedia Brittanica,* fifteenth edition, s.v. "ship."

7. E. L. Larrabee, The screw propeller, *Scientific American* 243 (1980), no. 1: 134–148.

8. R. E. Emerson, Design of a Man-Powered Boat, B.S.M.E. thesis, Massachusetts Institute of Technology, 1973.

9. M. B. Brewster, The Design and Development of a Man-Powered Hydrofoil, B.S.M.E. thesis, Massachusetts Institute of Technology, 1979.

10. S. F. Hoerner, *Fluid-Dynamic Drag* (Bricktown, N.J.: Hoerner, 1959).

11. J. C. McCullagh, *Pedal Power* (Emmaus, Pa.: Rodale, 177).

12. T. R. F. Nonweiler, Man-powered aircraft: A design study, *Journal of the Royal Aeronautical Society* 62 (1951): 723–734.

13. D. G. Wilson, *Human-Powered Space Transportation* (Boston: Galileo, 1978).

14. M. Shakespear, A Pedal-Powered Riding Lawn Mower, B.S. thesis, Massachusetts Institute of Technology, 1973.

15. J. F. Kincaid et al., The Automobile and Air Pollution: A Program for Progress, part II, report PB 176 885, U.S. Department of Commerce, 1967.

16. R. F. Post and S. F. Post, Flywheels, *Scientific American* 229 (December 1973): 17–23.

17. Report on the Energy-Storage Bicycle, Dartmouth College, Hanover, N.H., 1962.

18. F. R. Whitt, Freewheeling uphill—Is it possible?, *Cycling* (30 January 1965): 13.

19. Curry-Landskiff: Das neue verbesserte Modell 1930 (brochure), Alexander Metz, Munich.

20. How to build a rowmobile, *Science and Invention* (August 1926): 333.

21. W. Goepferich, Building a rowmobile, *Everyday Science and Mechanics* (January 1933): 130–132, 182.

22. A. K. Meier, Becaks, bemos, Lambros and productive pandemonium, *Technology Review* (January 1977): 56–63.

Recommended reading

History of Aviation, part 6 (London: New English Library, 1969).

Pedal-power flight beaten by wind, *Daily Telegraph* (London), 20 March 1972.

C. A. Marchos, Aerodynamics of Sailing (Adlard Coles: Granada, 1979).

Man-Powered Flight: The Channel Crossing and the Future (London: Royal Aeronautical Society, 1979).

H. Rouse, *Elementary Mechanics of Fluids* (London: Chapman and Hall, 1946), p. 286.

E. C. Shepard, What happened to man-powered flight, *New Scientist* (27 November 1969).

K. Sherwin, *Man-Powered Flight* (Hemel, England: Model and Allied, 1971).

Bicycling as a means of transport rose rapidly to an almost incredible level of popularity in the 1890s, as has been described in chapter 1. Many roads were either created or paved as a direct result of the Good Roads Movement of the League of American Wheelmen.[1] There was an outpouring of creative talent, and the designs of human-powered vehicles went through almost every possible variation before the combination of the pneumatic tire and the "safety-bicycle" configuration triumphed.

There have been very few changes in the design of the standard bicycle since 1890. Nearly "carbon-copy" bicycles, million after million, have been made since that time, with changes no greater than minor variations in wheel diameter, tire diameter, frame angles, and gear ratios.

America is a nation on wheels, with well over 100 million motor vehicles on the road. There are also over 100 million bicycles, as of 1981. Although one reason for this high figure is the affluence that makes it possible to buy a bicycle even if one does not intend to use it every day, it is still true that bicycling is the fastest-growing competitive and recreational sport in the United States. Many cities and states have designated bikeways, following the example of the initial one in Homestead, Florida. When the commissioner of parks closed Central Park in New York to all vehicles but bicycles on Sunday, the response of bicyclists was so great that it had to be concluded that a much larger portion of the population than is generally assumed would enjoy daily the gentle exercise of bicycling if it were not for the constant danger and unpleasantness of competing with automobiles for space on the roads.

That more Americans, even New Yorkers, can and will commute by bicycle when the pressures become high enough was shown during New York's two-week bus- and subway-workers' strike of 1980. City officials reported that 70,000 people were riding into Manhattan's business districts every day, with perhaps three times that number using bicycles citywide.[3] However, the number dropped when the strike was over. Furthermore, the separate bike lanes the city instituted in some places to encourage bicycling caused uncertainties and conflicts, and three (possibly four) pedestrians died after being hit by bicyclists.[4] These accidents illustrate the most serious impediment to increased future use of bicycles in the United States: With the exception of a few distinct communities (such as Davis, California and Homestead, Florida), state and local governments do not view bicycles as serious vehicles. Consequently, laws are haphazard. In the recent past some communities required bicyclists to ride on the right and some on the left, and in some places bicyclists were supposed to use the sidewalks, with little guidance on rules in pedestrian-bicyclist conflicts. The federal government has attempted to enforce uniform bicycle regulations, but these, along with almost all other road-use ordinances, are generally not enforced by the police or the courts. It is to the credit of the League of American Wheelmen that it has taken the leadership in endeavoring to educate bicyclists on the laws of road use and to have the laws enforced uniformly on bicyclists as well as on motor-vehicle users and pedestrians.

There is a great potential for increased bicycle use in the United States, and in most other countries with high levels of car ownership. The state of mild or severe anarchy on the roads is one of many contributing factors decreasing the attractiveness of this otherwise delightful way of getting around.

Most present commuting in the United States is accomplished by automobile. Most commutes are under 5 miles, and 36 percent are under 3 miles (figure 13.1). For trips of up to 5 miles, repeated trials of door-to-door journeys by ordinary bicyclists (not athletes) and automobile commuters have shown that the bicycle is on average the faster vehicle for urban use. The results shown in figure 13.2 are from a particular Sierra Club commuter race, but similar results have been obtained in many trials held in urban and dense suburban areas. Buses and subways are generally much slower when waiting times are included.

Automobiles are particularly inefficient for short trips. Figure 13.3 shows that on a 3-mile trip on a cool day, gasoline consumption will be about twice that normally quoted for the warmed-up car.[5] About 30 percent of all the gasoline consumed in the United States is used on trips of 3 miles or less.[6]

Although bicycles are clearly superior for many people and many trips, automobiles are used in the majority of cases. This is partly because they have been superbly engineered for convenience and comfort. However, the principal reason is that the use of automobiles is, despite current increases and complaints, extremely inexpensive. Gasoline is cheap by the standards of most of the world, and automobiles too are still inexpensive. Above all, use of the highway is inexpensive. All these factors are inexpensive because we have chosen to pay for the enormous costs the use of the automobile brings to society—principally through taxes. Our taxes increase while our overall standard of living decreases. It is the economic tragedy of the commons. Public transportation and bicycles cannot compete with this highly subsidized luxury. If automobile drivers were required to pay the approximately $4.00 per mile that economists calculate as the total external cost use of the car in an urban rush hour throws onto soci-

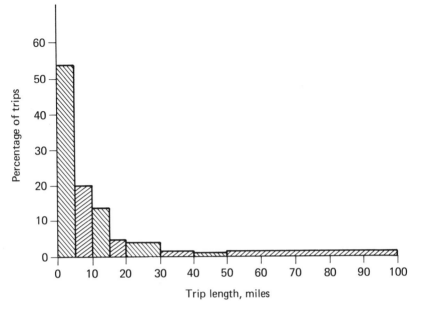

Figure 13.1
Percentage distribution of
automobile trips by
distance. Adapted from
*Automobile Facts and
Figures* (Detroit:
Automobile
Manufacturers'
Association, 1972).

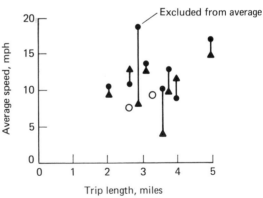

Figure 13.2
Results of Sierra Club
commuter race, May
1974 (●) Bicycle, (▲)
automobile, (○) tie.
Average speeds were 9.8
mph for automobiles and
10.8 mph for bicycles.

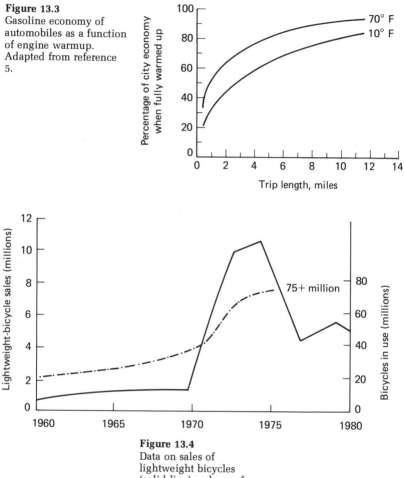

Figure 13.3
Gasoline economy of automobiles as a function of engine warmup. Adapted from reference 5.

Figure 13.4
Data on sales of lightweight bicycles (solid line) and use of bicycles (dashed line) in United States. Adapted from reference 2.

ety,[7,8] transportation would immediately become self-balancing—and more people would use bicycles for their own well-being and that of their community.

Figure 13.4 shows that more bicycles are now sold per year in the United States than automobiles. Of the more than 100 million bicycles owned, about 75 million are in use (though not that many are in regular use). In many cities around the world the proportion of person-trips by bicycle is substantial: 17 percent in Copenhagen; 20 percent in Uppsala, Sweden; 40 percent in Davis, California; 47 percent in Rotterdam; and around 45 percent in Stevenage, U.K. For general bicycle use to increase to these levels in the United States, economic and other disincentives (some of which were mentioned above and others of which will be touched on below) must be lessened or removed.

The bicycle is, in good weather and on smooth roads, an amazingly convenient means of transport. It gives instantly available door-to-door service, and at an average speed in urban areas that is usually better than that of any competitor (at least over distances of up to 5 miles, or 8 km, as mentioned above). It is extraordinarily light, with a payload of up to ten times the unladen weight, and it is narrow enough to travel through and be stored in places inaccessible to motor vehicles. A bicycle can pay for itself in saved transit fares in much less than a year. And, of course, it is an almost perfect way of getting exercise and keeping healthy.

All these attributes of this wonderful vehicle have been with us since before the turn of the century. So have nearly all of its shortcomings, some of which are listed here:

- The braking ability of all but specially equipped lightweight bicycles is very poor, especially in wet weather.

- A bicyclist without cumbersome clothing is unprotected from rain, snow, hail, and road dirt and from injury in minor accidents.

- It is difficult to carry packages, briefcases, shopping bags, and the like conveniently and safely.

- The aerodynamic drag in a headwind is very high.

- The riding position and the pedal-crank method of power input could be improved upon, ergonomically.

- Reliability is poor (especially that of brake and gear cables and wheel spokes), and in regard to maintenance the present design is attuned to the low labor costs of an earlier age.

- Whereas standard cars retail at about $2/lb ($4/kg), in 1981 prices, regular bicycles sell for about $4/lb ($8/kg) (and lightweight models may easily cost $20–200/lb), although bicycles contain much less sophisticated engineering than automobiles.

Correcting these drawbacks would provide little problem to NASA or General Motors. Their persistence is the consequence of a vicious circle that is similar to the one that has caused the running down of public transportation: Too many cars led to such unpleasant conditions for cycling that demand slackened; manufacturers cut out all "nonessential" expenditures; and nineteenth-century bicycles made poor competitors for highly developed modern automobiles. Having automobiles, affluent people moved farther and farther from the city centers to avoid the very congestion that automobiles had played a large part in causing. Commuting distances are now over 30 miles each way for many people—too far for all but the most dedicated bicyclist.

The situation may be changing. The unhappy state of our cities, the at-last-recognized harmful effects of automobile congestion in urban areas, the growing shortages of energy and raw materials, the concern over the damage to our environment—all of these are helping to recruit not only new bicyclists but also scientists and engineers anxious to solve problems.

Design competitions

It was hypothesized in chapter 1 that many of the developments in bicycle design in the last century were eagerly accepted only after they had proved their worth in competitions (usually speed races or time trials), and that the lack of notable developments since then has been partly due to the restrictions on the design of racing bicycles. However, innovative designs in components and in the overall vehicle have been increasingly in evidence in the last decade. These may have been spurred by three design competitions, and new thinking has certainly been greatly stimulated by the speed and time trials sponsored by the International Human-Powered-Vehicle Association (IHPVA) since 1974. We will discuss first the results of the design competitions.

The first modern international design competition was organized by D.G.W. and the journal *Engineering* in 1967–1968.[9,10] The aim of the competition was to encourage improvements in any aspect of human-powered land transportation. Let us look at the suggestions made by some of the 73 competitors, and at how their proposals would overcome some of the deficiencies, listed above, of present bicycles.

There were many proposals incorporating enclosures to give weather and minor-accident protection and luggage space, combined in some cases with a reduction in air drag. Some entrants recognized the penalties in increased weight, of side force in a cross wind, and of usually more difficult access. The bodies were virtually all added to a chassis or spine rather than being designed to supply structural strength; no one experimented with a "crustacean" rather than a "vertebrate" construction, and in this the competitors were probably wisely conservative. Whether the advantages given by an enclosure can justify its drawbacks will be known only through public acceptance. Most riders would not like to sacrifice the bicycle's narrow width and its ease of maneuvering and parking, but many would be well prepared to accept a weight penalty of 15 lb (6.8 kg) in a commuting vehicle if the body would keep the rider (and the briefcase or pocketbook) clean and dry, warm in winter, and as cool as possible in summer. The performances of some of the entries in the IHPVA speed trials, reviewed later, show that this penalty is justified.

Many competitors felt that it was logical to combine a body with a three- or four-wheeled configuration. Obviously there is an immediate addition of weight and of width for stability, if only because the wheels and suspension must now handle high side loads that are absent from bicycles. If we set out to attract a housewife (perhaps with a baby) to go shopping under her own power, we might find that a three-wheeler or a four-wheeler (which has one more wheel but one less track than the usual tricycle) would have a great appeal. The additional vehicle weight matters less when one is carrying cargo.

A configuration that might have advantages is that of a two-wheeled single-track vehicle with a semirecumbent riding position, a body, and outriggers that could be dropped when one stopped (figure 13.5). For a three-wheeler, the arrangement of a bicycle and sidecar gives two tracks instead of three and might have other advantages.

Body shape, rider attitude, and wheel arrangement are intimately connected with power transmission, and in this area competitors spent much creative effort. There was much preoccupation with constant-velocity foot motion in a straight line or through an arc, despite its apparent disadvantages.

Some entrants proposed hydrostatic transmissions, which would at least give efficient braking on the driven wheel and possibly a

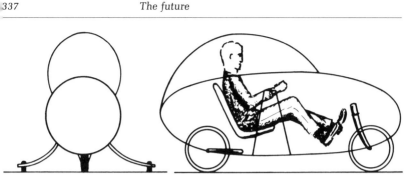

Figure 13.5
Design for enclosed
bicycle with outriggers.

continuously variable gear ratio. The weight
penalty, however, would be severe.

There was little evidence of emphasis on the
major problem of braking. The judges were disappointed at the lack of brake developments,
and would have given the first prize to anyone
who had made or modified a brake to give improved wet-weather operation and higher cable
reliability without adding greatly to the weight
or the cost.

The first prize went to W. G. Lydiard, who, besides carrying out careful design and analytical
work in the areas of stiffness, stability, aerodynamics, and transmission, made three experimental machines of different configurations. His
first model was a three-wheeler; the other two
(see figure 13.6) had two wheels 16 inches (406
mm) in diameter. Lydiard calls his Mark III machine (which he does not claim to be near a final solution) the Bicar, a name that correctly
implies that the rider is housed in a body and
pedals in a half-reclining position.

A problem identified by Lydiard with two-
wheeled reclining-rider bicycles is that either
the wheelbase and the overall length become
excessive or the feet must pedal over the front
wheel. He found that a conventional chainwheel
and cranks in this position gave a marked "feet-
up" attitude, and he eventually adopted pull
rods swinging through arcs in a more or less
conventional position. He found that these pull
rods interfered somewhat with his ability to put

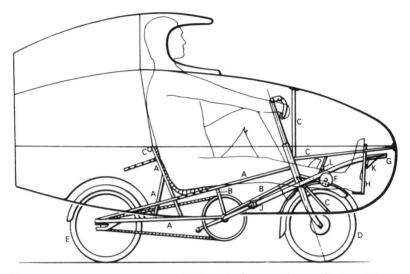

Figure 13.6
Lydiard Bicar Mark III, with half-reclining riding position. A: double-tube frame. B: single-tube frame. C: shell support. D: 16-inch front wheel with hub brake and generator. E: 16-inch rear wheel with hub gears. F: roller support for push rod. G: pull rod, actuating swinging crank. H: rocking pedal. J: universal joint. K: pedal stop. L: pull-rod bounce limiter. Rider puts legs through flaps in body to rest feet on ground. Towing tests indicated that this design could increase average touring speed by 6 mph. From reference 10.

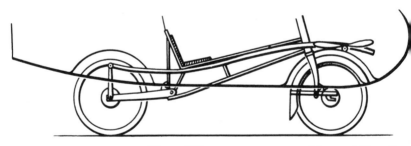

Figure 13.7
Lydiard Bicar Mark IV. Lydiard's proposed further development of his design would have sprung wheels, and possibly a variable-ratio friction gear in the rear wheel activated by the pull rods. From reference 10.

his feet on the ground through flaps in the body, and for a later machine (figure 13.7) he is considering pull rods operating a variable-ratio overrunning gear on the rear wheel, together with springing.

The Bicar's body is of 1-mm ABS plastic. Lydiard intended also to try $\frac{1}{2}$-mm ABS to reduce body weight, and also possibly $\frac{1}{4}$-inch (6.3-mm) paper honeycomb covered with Melanex, which would give an estimated weight of 5 lb (2.3 kg). He rejected, after some consideration, the idea of using the shell as the principal load-carrying member, and he employed a fairly conventional tubular spine frame. He decided to avoid the problems of windscreen fogging by leaving the rider's head in the open, on the theory that "no bicyclist would want to be hermetically sealed in, or object to the sun, wind or rain on his face in moderation."

From towing tests made to determine the drag, it was estimated that Lydiard's Bicar might increase a touring bicyclist's average speed (without stops) by up to 6 mph (2.68 m/sec). According to the calculations of chapter 7, this increment seems conservative.

Kazimierz Borkowski was another entrant who constructed a prototype. His machine (figure 13.8) is propelled by a sliding action of the seat along the long crossbar. The seat is attached to a carriage, which during the power (backward) stroke engages a long loop of chain coming from the rear wheel. The handlebars do not move longitudinally, so the rider must change position considerably during the stroke; this was of concern to the judges. Borkowski claims no more than that this is a "sport and recreation" vehicle that gives healthy exercise to more muscles than does normal cycling.

Stanislaw Garbien's vehicle (figure 13.9) was designed to transmit power to the rear wheel through swinging constant-velocity cranks and a continuously variable gear. This machine is a bicycle in which the rider sits fairly high over

Figure 13.8
Borkowski's rowing-
action bicycle. Power is
transmitted on the
backward stroke through
a sliding seat, which
runs on the long
crossbar. From reference
10.

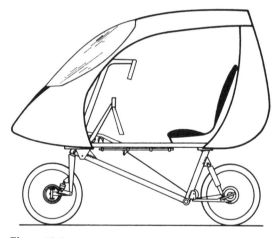

Figure 13.9
Garbien's semienclosed
bicycle design, with both
wheels sprung. Power is
transmitted to the rear
wheel through swinging
constant-velocity cranks
and an infinitely variable
gear. From reference 10.

the rear wheel and pushes levers over the front wheel. The open sides allow the rider to put feet on the ground when starting and stopping.

Early in 1974 the results of a design competition were published by the Japanese Design Promotion Organization and the Japan Bicycle Promotion Institute. The sponsors had advertised widely for entries, and had asked for "new ideas on cycle(s) which could be for any kind of vehicle with wheels and driven by human power." For comparative purposes the entries were categorized in three groups: "utility," "sport and pleasure," and "fantasy." The judges were disappointed at the lack of really new ideas. Prizes were awarded, however, because of the obvious hard work put in by the competitors.

The first prize was given to Terje Meyer, Bjorn Larsen, and Jan Christensen of Norway. Their multipurpose bicycle is shown in figure 13.10. They also entered other versions, including a tricycle. The judges reported that they were impressed by the team's ingenuity, both in the overall design and in the details. They gave much praise to the gearbox in the back hub, although it appears to be a version of the old constant-velocity system diagrammed in figure 2.13. They thought that the frame design had merits

Figure 13.10
Winning entry in Japanese Bicycle Promotion Institute design competition, 1974.

because it avoided the use of welded joints, a feature important for the use of aluminum as proposed by the designers. This design seems most ingenious, both in detail and overall, and worthy if novelty was the main aim of the exercise. It suffers from various drawbacks, however, when compared with conventional machines. For example, the frame, unless made relatively heavy through the use of a larger quantity of material to compensate for the long cantilevered members, would be flexible; the wheels would soon be twisted out of track. The long levers would be bent easily unless they were deep in section and therefore heavy. The nonpneumatic tires and small wheels would result in high rolling friction, giving probably a loss of 10–70 percent of the usual "utility" speed. Riding on rough roads would be uncomfortable, though perhaps the frame would flex somewhat and help to smooth bumps. The apparent lack of vertical adjustment to the saddle and handlebars seems another detriment. This lever design was claimed go give a "walking action," but with the lever ends pivoted at fixed points, giving largely vertical pedal motions, it appears that the foot motion would be better described as "stepping." Ergonomic testing has shown that, in terms of power output for a given oxygen-breathing rate, stepping exercise is not superior to that of pedaling on a conventional crank system. Cycling history shows that as soon as good chain drives became available (in the 1880s) the rotary pedaling system was generally accepted. Lever systems have reappeared spasmodically as "innovations," but no physiological experiments have backed up the innovators' enthusiasms. The advent of reliable variable gearing for bicycles probably gave the first place to rotary drive in that they became as acceptable for low road speeds uphill as lever drives, which had been claimed to be better.

The second prize in the Japanese design com-

Figure 13.11
Second-prize entry in
Japanese Bicycle
Promotion Institute
design competition,
1974.

petition was awarded to the Japanese designers
of two folding bicycles. Figure 13.11 shows one
of these designs, which when folded would be
easy to carry. The designs incorporated short
wheelbases, small wheels, and dubious means
of altering saddle height. The short wheelbase
seems hazardous, and the dimensions of the
opened-up machines seem unsafe for other than
flat roads. The steering would be very sensitive,
and except on hard, smooth, flat pavements the
high rolling resistance would be a disadvantage
for other than very short distances. In general,
folding bicycles are intended for short trips
(from train station to office, for example); perhaps these designs served this purpose.

Another prize was awarded to Naef Fridolin of
Switzerland for the "fun" or "fantasy" machine
shown in figure 13.12. The rider sits inside a
spokeless 1.7-m-diameter wheel and steers via a
trailing wheel. The report says that the testers
experienced peculiar sensations. (We remarked
on the difficulties of rear-wheel steering in
chapter 9.) Curiously, similar machines appeared in the last century and were proposed as
serious competitors for the less outlandish bicycles and tricycles.[11]

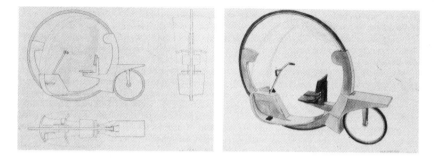

Figure 13.12
Third-prize entry in
Japanese Bicycle
Promotion Institute
design competition,
1974.

The Japanese competition attracted no designs
for racing bicycles and offered nothing for serious cycle tourists. The most practical entries appear to have been generated mainly by a need
for a machine easily stowed in a car trunk and
meant for traveling short distances on city
streets.

The British Cycling Bureau, sponsored by
manufacturers and dealers, joined with the Sunday *Times* magazine and the *Observer* to organize a bicycle-design competition for permanent
residents of the United Kingdom. The competition had three categories: working cycles (machines for short local trips, shopping, and short-
distance commuting), leisure cycles (machines
for longer-distance leisure cycling), and accessories. The three winning entries (announced in
April 1979) for a "working cycle" all had small
wheels, were foldable, and emphasized the convenience of the bicycle for about-town transport.
The first prize when to Juan Szumowski, a London architect, for a bicycle that folded into a
hand-carried package (figure 13.13). The shaft-
drive transmission was to be made largely of
molded plastics. The second prize went to the
WW group, whose design concealed a rein-
forced-rubber-belt transmission in a molded-
plastic box frame (figure 13.14). This design
also included a "power point" for recharging
light batteries, a built-in cable lock, and saddle-
height adjustment by means of a compressed
gas. These two entries were beautifully made

Figure 13.13
Winning entry in British
Cycling Bureau design
competition, 1979.
Courtesy of British
Cycling Bureau.

but nonfunctional mockups. Grahame Herbert's
third-prize entry was a functional folding bicycle of conventional materials (figure 13.15). Despite an enthusiastic entry of nearly 100
designs, the juding panel reluctantly decided
not to award prizes in this leisure-cycle category. On the whole, competitors in this category
either added design details to conventional
cycles without enlarging their function and
range or submitted imaginative ideas that
proved on examination to be unworkable.

The British competition, like the previous one
in Japan, showed the present interest in foldable
bicycles. The judges chose, for the premier
awards, entries that emphasized style rather
than function. If the plastic machines were "engineered" to be sufficiently stiff to be pedaled
efficiently, and with stresses low enough to
avoid fatigue failures, they would inevitably be
heavier than more conventional present ma-

Figure 13.14
Second-prize entry in
British Cycling Bureau
design competition,
1979. Courtesy of British
Cycling Bureau.

Figure 13.15
Third-prize entry in
British Cycling Bureau
design competition,
1979. Courtesy of British
Cycling Bureau.

chines. However, the previous two design competitions led to trends or to actual commercial production (by, for instance, an Italian manufacturer of an ingenious lever-drive bicycle very similar to that of figure 13.10), so we may see efforts to satisfy the evident desires of part of the public for high-style easy-care bicycles like those of figures 13.13 and 13.14.

IHPVA races

The International Human-Powered Vehicle Association (IHPVA) was formed in 1974 by Chester Kyle and fellow enthusiasts, principally as a body through which the record speeds reached on the bicycles they were building could be recognized. The only restriction on the design of the vehicles which could be entered in their annual speed trials was that there be no energy storage other than that in the riders. The existing organization that oversaw bicycle racing had extremely tight design restrictions. We argued above that these restrictions had helped to stifle developments in bicycle design for much of this century. Whether or not this is true, there is no doubt that the IHPVA speed trials (and, more lately, the one-hour distance race) have had and are having a dramatic effect on human-powered racing vehicles. We shall discuss later the question of the influence of these special vehicles on commuting and recreational bicycles.

The early (1974) IHPVA records (figure 4.9) were set by standard bicycles with faired enclosures. One designed by Kyle himself had the extremely low drag coefficient of 0.10 (table 4.3). However, in essence these vehicles were similar to others that had been made at intervals throughout this century—notably, perhaps, those of Oscar Egg in Britain (figure 4.5). These former models were merely curiosities, because any records they set were not recognized by bicycling's governing body, but the new IHPVA machines and their recognized records started an inexorable process of development and refinement.

In all the later successful models, the frontal area was reduced below that of a faired standard racing bicycle plus rider. To do this required that the rider in a single-rider vehicle lie prone or supine. Allan Abbott constructed a bicycle with an overhead girderlike frame connecting front and rear wheels, and suspended himself face down horizontally beneath it (figure 13.16). Abbott had already set a motor-paced speed record of 138.67 mph (62 m/sec) on a more conventional bicycle. When he used a fairing with the prone recumbent (figure 13.17), he became the fastest self-propelled human being. However, control was difficult, possibly because of the flexible connection between rider and vehicle. Most later machines were designed to have the riders supine, or else prone and supported on seats or pads rather than in slings to give a much stiffer connection and better handling.

The 50-, 55-, and 60-mph "barriers" were broken by a tricycle called *White Lightning* (figure

Figure 13.16
Abbot's prone recumbent bicycle, with crude body. Courtesy of Chester Kyle.

Figure 13.17
Abbot's prone recumbent
bicycle with fairing.
Courtesy of Chester Kyle.

13.18) constructed by students from Northrup
University and pedaled by two supine riders,
one behind the other. This vehicle was beaten
in 1979 by a machine, built by a team called
Vector, that was thought by some to approach
the ultimate in low drag and high power output. It was a quadracycle with three prone
riders arranged head to feet (figure 13.19). The
front rider steered with his hands and pedaled
normally; the second and third riders turned
handgrips on the pedals of the rider ahead in
addition to the pedaling. It is estimated, as
stated in chapter 2, that about 12.5 percent more
power can be delivered if the hands are used as
well as the legs, especially in short-duration
maximum-power attempts.

Despite their 1979 success, the Vector design
team (led by Al Voight) produced two quite different vehicles for the 1980 speed trials (figure
13.20). The riders were semireclining in each,
with their heads up; thus, they were able to see
between their knees. In the two-person Vector,
which reached 62.92 mph (28.13 m/sec), the
riders sat back to back and used pedals only.
The single-rider Vector reached 56.66 mph
(25.33 m/sec), faster than any multiple-rider vehicle had gone until a year before. Both were
tricycles with the single driving wheel in the
rear, behind the rider's back in the single-rider

Figure 13.18
The supine tandem
tricycle *White Lightning*.
Courtesy of Chester Kyle.

Figure 13.19
Vector three-rider
quadracycle. Courtesy of
Chester Kyle.

Figure 13.20
Vector semirecumbent
tricycles, 1980. Courtesy
of Chester Kyle.

case. The transmission was a fairly conventional
six-speed derailleur, with added "idler" sprockets to route the long chain beneath the rider(s).
The 100-tooth chainwheel, in combination with
an 11-tooth top-gear sprocket, required only 88
rpm on the cranks to produce 60 mph. This
supports the Japanese data reproduced in chapter 2 showing crank rpm between 60 and 100
giving maximum output and efficiency. (Much
higher rpm are frequently advocated.)

Some data on the two 1980 Vector vehicles are
shown in table 13.1. When these are compared
with the hypothetical values listed for the "ultimate human-powered vehicle" in table 7.2,
some close agreements can be seen. The rolling
resistance of the Vectors seems to be above the
minimum predicted from Kyle's tests,[12] so there
should be some scope for further development.

Table 13.1 Data on Vector vehicles, 1980.

	Single-rider	Tandem
Length	2.95 m	3.84 m
Width	0.635 m	0.635 m
Height	0.813 m	0.838 m
Weight	23.13 kg	34.02 kg
Frontal area	0.424 m^2	0.437 m^2
Rolling-friction coefficient	0.006	0.005
Aerodynamic-drag coefficient	0.11	0.13

The frontal area will be marginally reduced, no doubt, but there does not seem to be much prospect for significant reductions in the aerodynamic drag coefficient unless some sophisticated forms of boundary-layer control are used. Therefore, the principal scope for even higher speeds seems to lie in delivering more power to the driving wheel, which can be achieved only through the following:

• riders with higher power-weight ratios,

• longer acclimatization and training,

• reduction in transmission losses,

• use of arm and possibly trunk muscles in addition to leg muscles, and

• use of motions of feet, hands, etc., by which more power is given by the human body.

Future commuting vehicles

Developments in racing or speed-trial vehicles will influence the design of commuting vehicles, but the two types of machines will be far from identical as long as human-powered commuting vehicles have to share the roads with motor vehicles. Commuting vehicles must be visible to motor-vehicle drivers, some of whom sit 6 feet or more above the ground in trucks of over 20,000 pounds. A flag on a tall stalk makes a conventional bicycle more visible, but a flag alone, no matter how large, will be insufficient to compensate for the dangers of being in a vehicle less than a meter high in heavy traffic.

Therefore, a commuting vehicle should be somewhat higher than a speed-trial machine.

An appropriate rider height is that used in present automobiles, with the seat about 2 feet (0.6 m) above the ground. The recumbent position gives lower frontal area, and the supine position is more comfortable and natural than the prone. Also, it is safer to hit an object with the feet than with the head. Although the 1980 Vectors proved the semisupine position to be ergonomically effective, a higher seat-back angle would be desirable to give a greater field of vision as well as more visibility.

An enclosure is essential to reduce aerodynamic drag and to give protection from rain and snow. Throughflow ventilation that can be brought into operation in hot weather will also be essential. We do not yet know to what extent an enclosure will prove detrimental at extremely high temperatures. An enclosure made of the 10-mm styrofoam used for inexpensive insulated picnic boxes would provide excellent thermal insulation and some injury protection. Present materials of this type are easily damaged, and stores would need to stock replacement fairings (which, in high volume, should cost no more than a pair of tires). It would be practicable to incorporate an automobile-type "radiator" (actually a convection heat exchanger) that could be filled with crushed ice at the start of short commuting trips in hot weather and would cool the inflowing ventilation air. For longer trips in extreme heat, riders would have to rely on maximum ventilation and, as at present, reduced exertion.

All recent winners of the IHPVA speed trials have had more than two wheels. This has led many writers to conclude the future commuting vehicles (dubbed somewhat imprecisely "phase-3 bikes") would not, in fact, be bicycles. But tricycles and quadracycles have serious weight and size disadvantages. If a pair of wheels is used for steering, an Ackermann-type steering linkage must be used. If two wheels are drivers, some form of differential has to be incorporated. All wheels and their supporting structures have to be designed to take the large side loads resulting from, for example, a sudden high-speed turn of a heavily laden machine at the bottom of a long hill. (The wheels of a bicycle must withstand forces only in the plane of the wheel.) The wheel track also has to be wide enough for the vehicle not to overturn in the same type of maneuver, depositing its rider into the path of oncoming traffic. The width of a multiwheel vehicle would force it to occupy a full traffic lane, incurring the constant wrath of motor-vehicle users, whereas a bicycle can comfortably use a half lane or less. There have been many attempts to introduce three- and four-wheeled pedaled vehicles for highway use (see figure 12.16). We believe that their failures were due more to their widths and to the weight required to make them sufficiently stiff than to any other reason.

We believe that future human-powered commuting vehicles will be bicycles. They may incorporate outrigger wheels such as those shown in figure 13.5 to be used at rest and during startup. They will have enclosures or fairings, but these will be shorter than the very-low-drag shapes used for speed-trial vehicles, for the sake of maneuverability, side-force reduction in crosswinds, and easier storage. The riding position will be semirecumbent. The hands and arms will be at the rider's sides (so that potentially injurious steering gear can be removed from in front), and may be used for power as well as steering. There will be built-in luggage capacity and theft-resistant lights. These vehicles will be heavier than present commuting bicycles, but their flat-road performance will be very attractive (see figure 7.4 and table 7.2).

The additional weight will make these vehicles less attractive in very hilly regions. One obvious approach is to dispense with the fairing in such

Figure 13.21
The Syracuse Crusway
powered-guideway
concept. Courtesy of
Syracuse University
Research Corporation.

areas. A fairing would give a measurable drag
reduction only on the downhills, where one
would like more drag, and would increase
weight (although one hopes the increase would
be small) and thus add to the effort required on
the uphills. Another approach, investigated at
Syracuse University, is some form of off-vehicle
powered assistance for steep hills. The Syracuse
"Crusway" (figure 13.21) had a form of ski-slope
T bar coming from overhead and hooking onto
the handlebars of conventional bicycles. Reference 10 proposed a moving belt in the roadway
onto which one would pedal one's tricycle, with
the single front wheel on the belt, apply the
front brake, and accelerate to belt speed. This
could work for bicycles with deployable outrigger wheels. Another possibility, which would
be feasible for regular bicycles, would be to
have a moving handrail like those on escalators

Figure 13.22
Powered handrail for
uphill assistance.

(figure 13.22). Powered assistance systems
would preferably be installed in a separate
right-of-way, although the moving handrail
could be used on ordinary streets if parking
were prohibited on the uphill curb. The handrail could also be used by skaters on the
sidewalk.

A saner future

Whether any of these seemingly desirable developments will actually take place, or whether the
world will continue to rush to utilize every new
discovery of stored energy in ever-more-extrava-
gant "power trips," cannot be predicted. What
can be forecast is that the former pattern of dou-
bling energy consumption every decade or so
cannot continue much longer, for many reasons
of which the limited availability of energy is
only one. Pollution, land-use problems, and the
question of where to get materials from which
to make all the energy-using gadgets this increasing consumption presupposes are almost
immediate problems in several countries. Mankind's energy dissipation, now about 1/20,000
of the incident solar energy, would reach the

level of the sun's radiation to the earth in about 110 years if we continued the present rate of increase. Obviously, long before this could occur the climate would be so modified as to make irreversible changes in the whole of the earth's ecology, and life would probably be impossible for many plants and animals.

The gentle way of the bicycle is, for short distances, a transportation alternative that is compatible with nature and with a way of life that many of us would find an improvement over today's frenetic rushing hither and thither. We believe that the present renewed enthusiasm for bicycling is an encouraging sign of a saner future.

References

1. Bates, Good common roads and how to make them, *The Wheelmen* (Boston) V (1885): 194–200.

2. A calendar of the '70s, *American Bicyclist* 100 (1979), no. 12: 261.

3. J. Krohe, Jr., America on wheels, *Across the Board* (June 1980): 49–57.

4. W. K. Ezell, New York bike lanes: Use them or lose them, *American Wheelmen* (January 1981): 4–9, 20.

5. T. C. Austin and K. H. Hellman, Passenger-Car Fuel Economy as Influenced by Trip Length, paper, Society of Automotive Engineers, 1975.

6. R. A. Rice, System Energy as a Factor in Considering Future Transportation, paper, Society of Automotive Engineers, 1970.

7. R. J. Smeed, *Road Pricing: The Economic and Technical Possibilities* (London: HMSO, 1964).

8. G. Roth, *Paying for Roads* (Harmondsworth, U.K.: Penguin, 1967).

9. D. G. Wilson, A plan to encourage improvements in man-powered transit, *Engineering* (London) 204 (1967), no. 5283: 97–98.

10. D. G. Wilson, Man-powered land transport, *Engineering* (London) 2071 (1969), no. 5372: 567–573.

11. A. Ritchie, *King of the Road* (Berkeley, Calif.: Ten-Speed, 1975), pp. 48–49.

12. C. R. Kyle and W. E. Edelman, Man-Powered-Vehicle Design Criteria, Third International Conference on Vehicle Dynamics, Blacksburg, Va., 1974.

Appendix

Conversion factors

Mass: x lbm = 0.4536x kg (kilograms)

Force: x lbf = 4.448x N (newtons)

Length: x in. = 25.4x mm (millimeters)
x ft = 0.3048x m (meters)
x miles = 1.609x km (kilometers)

Area: x ft² = 0.0929x m²

Volume: x ft³ = 0.02832x m³

Pressure, stress, modulus of elasticity: x lbf/in.² = 6,895 Pa (pascals)
         (1 Pa = 1 N/m²)
        = 6.895x kPa (kilopascals)
(100 kPa = 1 bar = 14.503 lbf/in.²)

Density: x lbm/ft³ = 16.017x kg/m³

Velocity: x mph = 0.447x m/sec
         (meters/second)
        = 1.609x km/h
         (kilometers/hour)
x knots = 0.52x m/sec

Torque: x lbf-ft = 1.356x N-m (newton-meters)

Energy: x ft-lbf = 1.356x J (joules)
x Btu = 1,054.9x J
x kcal = 4,186.8x J
x kWh = (3.6 × 10⁶)x J = 3.6 MJ
        (megajoules)

Power: x hp = 746x J/sec = 746x W (watts)
x kcal/min = 69.78x W
x ft-lbf/sec = 1.356 W

Specific heat: x Btu/lbm-°R = 4,187x J/kg-°K

Heat flux: x Btu/ft²-h = 3.154x W/m²
x kcal/m²-h = 1.163x W/m²

Derivations

Force (newtons) = Mass (kilograms)
                  × Acceleration (m/sec²)

Energy or Work (joules) = Force (newtons)
                          × Distance (m)

Power (watts) = Work (joules)
                per Unit time (seconds)

Mass and weight

When we refer to the weight of (for instance) a bicycle or its rider, we are, strictly, giving the gravitational force. The correct units would therefore be newtons or pounds force (lbf). If we were to take a bicycle to the moon, its weight would be about one-sixth of its weight on the earth. The mass would remain unchanged. Therefore, we have usually given the mass (in kilograms or in pounds mass, lbm) when we have by common usage referred to the "weight." Weight is given by the relation

$$\frac{\text{Mass} \times \text{Gravitational acceleration}}{g_c},$$

where g_c is a constant that in the S.I. system equals unity and in English units equals 32.17 lbm-ft/lbf-sec².

**Properties of dry air
at normal pressures**

Temperature			Specific heat C_p (kJ/kg-°K)	Thermal conductivity k (kW/m-°K)	Density[a] ρ (kg/m³)	Viscosity[a] (m²/sec)
°K	°C	°F				
275	2	35.6	1.0038	2.428×10^{-5}	1.284	1.343×10^{-5}
300	27	80.6	1.0049	2.624×10^{-5}	1.177	1.567×10^{-5}
325	52	125.6	1.0063	2.816×10^{-5}	1.086	1.807×10^{-5}

a. These properties are at 1 bar, atmospheric pressure.

**Gear-speed
conversion chart**

Note: This chart is derived from one issued by the Tandem Club (U.K.). In continental Europe, the gear size is often specified as $\pi \times$ meters, or $3.1416 \times$ meters, which gives the distance traveled for one revolution of the cranks.

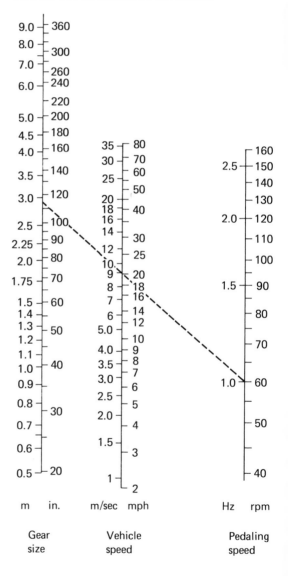

Index